Computational Biology

Volume 18

The *Computational Biology* series publishes the very latest, high-quality research devoted to specific issues in computer-assisted analysis of biological data. The main emphasis is on current scientific developments and innovative techniques in computational biology (bioinformatics), bringing to light methods from mathematics, statistics and computer science that directly address biological problems currently under investigation.

The series offers publications that present the state-of-the-art regarding the problems in question; show computational biology/bioinformatics methods at work; and finally discuss anticipated demands regarding developments in future methodology. Titles can range from focused monographs, to undergraduate and graduate textbooks, and professional text/reference works.

More information about this series at http://www.springer.com/series/5769

Florian Frommlet · Małgorzata Bogdan
David Ramsey

Phenotypes and Genotypes

The Search for Influential Genes

 Springer

Florian Frommlet
Center for Medical Statistics, Informatics,
    and Intelligent Systems
Section for Medical Statistics
Medical University of Vienna
Vienna
Austria

David Ramsey
Department of Operations Research
Wrocław University of Technology
Wrocław
Poland

Małgorzata Bogdan
Institute of Mathematics
University of Wrocław
Wrocław
Poland

ISSN 1568-2684
Computational Biology
ISBN 978-1-4471-7380-9   ISBN 978-1-4471-5310-8   (eBook)
DOI 10.1007/978-1-4471-5310-8

Preface

In the past 20 years we have witnessed revolutionary technological development in the fields of biology/genetics and computing. This has enabled the success of the Human Genome Project and the sequencing of a huge proportion of the human genome. However, this achievement has not reduced the number of questions related to the influence of genes on a multitude of traits and the general well-being of living organisms, although the availability of new tools has enabled us to identify complicated genetic mechanisms, such as DNA methylation or gene–gene regulation.

The systematic increase in the availability of good quality genetic data has aided efforts toward a more complete description of the genetic background of complex traits, i.e., those that are determined by many genes, often interacting with each other. Research in this area is rapidly expanding, since, apart from extending knowledge in the field of biology, it addresses many socially/economically important problems. Marker (gene) assisted selection is currently widely applied to identify promising individuals for breeding programs among domesticated animals, leading to increased efficiency in production or enhancing the quality of food products such as milk or meat. In the context of human genetics, the identification of influential genes allows us to evaluate an individual's susceptibility to certain diseases, design tools for early diagnosis, and produce new efficient medicines or personalized therapies.

As a result of this technological breakthrough, bioinformatics has appeared as a new scientific discipline, where the most effective research is performed by collaboration between biologists, computer scientists, and statisticians. While search through large and rapidly expanding genetic databases enables the identification of new genetic effects, it also creates a multitude of computational and statistical problems. Concerning statistical issues, the large dimension of statistical data often results in an erroneous description of reality when oversimplified statistical tools are used for their analysis. A full understanding of the properties of statistical/bioinformatics methods in such a high-dimensional setting is needed to accelerate progress in this field and requires further intensive research.

Understanding the properties of various methods for analyzing high-dimensional data requires advanced mathematical tools, while the development of efficient computational methods requires advanced knowledge in computer science. Therefore, the main intended audience of this book is students/researchers with a background in mathematics or computer science, who would like to learn about problems in the field of statistical genetics and statistical issues related to the analysis of high-dimensional data. Thus, we expect that readers possess some mathematical or computer science skills. On the other hand, the genetic material is explained starting at a basic level. For those who are not totally familiar with the fundamentals of statistics, an extensive statistical appendix is presented for reference.

While bioinformatics and statistical genetics deal with a variety of complex questions in the field of genetics, in this book we concentrate on methods for locating influential genes. Thus, we mainly discuss methods of identifying the associations between the *genotypes* of genetic markers and interesting traits (*phenotypes*). Also, we do not discuss methods based on pedigree analysis or family relationships, often applied in studies on humans or domesticated animals. Instead, we cover in detail methods of gene mapping in experimental crosses, as well as genome wide association studies, which are based on a random sample of individuals from outbred populations (e.g., from general human populations). We summarize classical and modern methods for gene mapping and point toward related statistical and computational challenges. We believe that the knowledge contained in our monograph forms an excellent starting point for becoming involved in the exciting world of this field of research and hope that at least some of our readers decide to take this invitation and participate in the ongoing journey to develop a better understanding of the role of genetics in the biology of living organisms.

Vienna
Wrocław
August 2015

Florian Frommlet
Małgorzata Bogdan
David Ramsey

Contents

Acronyms

ABOS	Asymptotic Bayes optimality under sparsity
AIC	Akaike's information criterion
BH	Benjamini Hochberg procedure
BIC	Bayesian information criterion
BLUP	Best linear unbiased predictor
CAT	Cochran Armitage trend test
CDCV	Common disease—common variant
CDRV	Common disease—rare variant
CEU	HapMap Population: Utah, Europe ancestry
CHB	HapMap Population: Han Chinese in Beijing
CNV	Copy number variations
DNA	Deoxyribonucleic acid
EBIC	Extended Bayesian information criterion
FDR	False discovery rate
FWER	Family wise error rate
gFWER	Generalized family wise error rate
GLM	General linear model
gLM	Generalized linear model
GWAS	Genome wide association study
HWE	Hardy-Weinberg equilibrium
IBD	Identical by descent
JPT	HapMap population: Japanese in Tokyo
KL	Kullback-Leibler
LASSO	Least absolute shrinkage and selection operator
LD	Linkage disequilibrium
LMM	Linear mixed model
LRT	Likelihood ratio test
mBIC	A modification of the Bayesian information criterion
ML	Maximum likelihood
MOSGWA	Model selection for genome wide association (software)
mRNA	Messenger RNA

MTP	Multiple testing procedure
NEG	Normal exponential gamma distribution
PCA	Principal component analysis
PCER	Per-comparison error rate
PFER	Per-family error rate
QTL	Quantitative trait locus
REML	Restricted maximum likelihood
RNA	Ribonucleic acid
SD	FDR controlling step-down procedure
SIS	Sure independence screening
SNP	Single nucleotide polymorphisms
YRI	HapMap population: Yoruba in Ibadan

Chapter 1
Introduction

The advances in the field of genetics over the past two generations have been astounding. The double helix structure of DNA, the genetic basis of life and reproduction in humans and many other species, was first described in print in 1953 [1]. In the little over 60 years that have passed since then, we have developed genome sequencers which can read the whole human genome composed of approximately 3.2 billion nucleotide bases. On the one hand, these advances have enabled us to answer questions regarding the evolutionary relationship between species and given us a greater understanding of a large number of diseases which have a genetic source, e.g., cystic fibrosis [2]. On the other hand, this rapid development has raised many new questions to be answered. Many diseases have both genetic and environmental factors, in particular cancers [3]. In such cases, the mechanisms underlying the susceptibility of an individual to a condition and triggers determining whether, and if so when, such an individual will develop that condition often involve a network of genes, as well as environmental effects [4].

Such problems are by their nature interdisciplinary. Communication and cooperation between scientists are required even to start answering many of these questions. Insights from geneticists and bioinformaticians continue to be necessary to develop the technological software which is now available. None of these advances would have been possible without the incredible acceleration in computing speed and memory. Insights from geneticists and statisticians are necessary to build models. Bioinformaticians and statisticians are needed to analyze data, but require geneticists and biologists to explain the mechanisms underlying the patterns seen in the data. Many recent advances in statistical theory have been in response to the emergence of "big data", i.e., huge data sets, in particular genetic data. However, as always, these advances are part of the continuing journey that underlies scientific progress and we are far from understanding many of the issues presented by such data sets.

This book aims to be a guidebook to part of this journey. Specifically, we look at developments in the studies of genetic association. The title of the book "Phenotypes and Genotypes" reflects this. Studies of genetic association aim to elucidate how our genetic code (genotypes) influence the traits we possess (phenotypes). This a relatively new and rapidly expanding field. Overall, the aims of the book

© Springer-Verlag London 2016
F. Frommlet et al., *Phenotypes and Genotypes*, Computational Biology 18,
DOI 10.1007/978-1-4471-5310-8_1

are to present the theoretical background to studies of genetic association (both genetic and statistical), indicate how the field has advanced in recent years, give a snapshot of the most commonly used methods at present, together with their advantages and shortcomings, and finally indicate some of the problems that remain to be solved in the future. Since the authors are statisticians, stress will naturally be placed on the statistical models and methods involved. But by necessity, in order to understand the statistical models, one must first understand the biological concepts underlying the statistical models.

More specifically, Chap. 2 gives an overview of the concepts from genetics required to be able to interpret and develop statistical models of genetic association. The ideas of phenotype and genotype are fundamental. A phenotype is any observable trait of an organism. In particular, here we will be interested in dichotomous traits (i.e., only two states are possible, for example, the presence or absence of a disease) and continuous traits (these are traits which are measured according to some scale, e.g., height, weight, milk yield).

We then give an overview of the genome. This is the genetic information which is found in each cell of an organism. The theory will concentrate on diploid organisms. Genetic information in such organisms is contained in pairs of chromosomes, humans have 23 such pairs. One chromosome of each pair is inherited from an individual's mother and the other comes from the father. In many organisms, including humans, one of these pairs is associated with the sex of an individual. The other pairs of chromosomes are called homologous, since the genetic information found at a pair of corresponding loci on such chromosomes combines to form an individual's genotype. In practice, we observe the genotype of an individual at a given locus, but we do not know which information came from the mother and which from the father.

Suppose for simplicity that two simple traits, say eye color and blood group, are each coded by a single gene. If these genes are located on different chromosome pairs, then the information passed on by a parent regarding one trait is independent of the information passed on regarding the second trait (in each case the information comes from the maternal chromosome with probability 0.5 and otherwise comes from the paternal chromosome). However, if the genes for these two traits are located close to each other on the same chromosome pair, then it is likely that the information passed on by a parent very likely comes from the same chromosome (either the maternal or the paternal). In this case, the genes for these traits are said to be linked, or equivalently that the two corresponding loci are linked. We consider the genetic distance between two loci, whose definition is based on the probability that the information passed on by a parent at two loci comes from the same chromosome. This is one minus the probability of a so-called crossover, which occurs when the information passed on at two loci on the same chromosome originally came from different chromosomes. The possibility of crossover results from the recombination of genetic material on homologous chromosomes before it is passed on to offspring. The closer two loci are on a chromosome, the less likely crossover is. Some probabilistic models linking genetic distance to the actual physical distance between loci are presented.

In general, the relation between traits (phenotypes) and genotypes is far more complex than the determination of eye color. For example, sex obviously has an influence on the height of humans, but the height of individuals of a particular sex follows a normal distribution. From the Central Limit Theorem, it would seem that height is affected by a large number of factors. Studies have shown that height depends on both environmental factors and various genetic loci [5]. For species with a short life span, experimental populations have been produced by the associative breeding of lines, where within each line individuals share a (distinct) set of simple traits and differing values of a given quantitative trait, e.g., one line of tall individuals and one line of short individuals. We know the location of many genes which define simple traits. Such loci are called markers. By appropriately crossing inbred lines, it is possible to create experimental populations, which can be used to discover associations between simple traits and the value of quantitative traits. These methods: backcross, intercross, and recombinant inbred lines are also discussed in Sect. 2.2.1. Suppose, as a result of such an experiment, a quantitative trait is strongly associated with a simple trait. This indicates that a gene that strongly affects the quantitative trait is located very close to the gene responsible for the simple trait. Such problems and statistical methods for locating quantitative trait loci (QTL mapping) are considered in more detail in Chap. 4.

Obviously, in the case of many species, particularly humans, it is impossible to create such experimental populations. However, the emergence of genome wide sequencers has led to the possibility of carrying out so-called Genome Wide Association Studies (GWAS). The general concepts behind the design of such studies are outlined in Sect. 2.2.3. In such studies, the number of genetic variables considered is generally much greater than in QTL mapping, and so the statistical problem of multiple testing becomes much more serious. This problem arises from the fact that applying classical procedures of hypothesis testing, i.e., using a fixed significance level, very often leads to a large number of false discoveries.

It should be noted that the classical probability and statistical theory, which form the basis for Chaps. 3–5, are described in the Appendix. Readers who are not familiar with this theory should first read the Appendix, before proceeding to Chaps. 3–5. Other readers should use the Appendix as a source of reference when necessary.

Chapter 3 is split into two main sections. Section 3.1 describes statistical approaches to solving the multiple testing problem and the relationship between such procedures and Bayesian decision theory. Section 3.2 deals with methods of model selection.

Consider a simple situation in which we have m markers on one chromosome and we wish to test whether there is a QTL on the same chromosome. In order to do this, we might carry out a set of m tests where the null hypothesis of the i-th test states that the i-th marker is not associated with the quantitative trait in question and the alternative is that the i-th marker is associated with the quantitative trait. One might carry out all these tests at a significance level of 5 % and conclude that there exists a QTL on the same chromosome if and only if the null hypothesis is rejected at least once. One obvious problem with this approach is that as m increases, the probability of accepting that there is a QTL on the same chromosome also increases. In such a case, controlling the familywise error rate (FWER) rate is an appropriate criterion to ensure that the probability of any false detection (i.e., concluding there is a QTL on that chromosome, when there is none) remains low, regardless of the number of markers used. The classical approach to this problem would be to use the Bonferroni procedure, which involves dividing the nominal significance level (here 5 %) by the number of markers. This ensures that the FWER does not exceed 5 %. Refinements of the Bonferroni procedure are also considered.

When the number of tests used is very large, procedures based on the Bonferroni procedure tend to be very conservative, i.e., for m large, very often we fail to detect a real association. This is particularly crucial when the goal is not to test an overriding hypothesis (i.e., the hypothesis that there is no QTL on chromosome versus the alternative that there is a QTL), but to discover which loci are associated with a given trait (i.e., the individual hypotheses are important in themselves). In such cases, an appropriate criterion for multiple testing procedures is to control the false discovery rate (FDR). Use of such a procedure ensures that the expected proportion of discovered associations that are not real associations is at most 100α %. The Benjamini–Hochberg (BH) procedure controls the FDR. The BH procedure is also less conservative than the Bonferroni procedure, particularly when there are a large number of real associations, and thus detects real associations more often than the Bonferroni procedure.

In general, we expect that only a small proportion of loci are real factors in determining a trait. Such cases are known as sparse. Bayesian decision theory can be applied to such problems by assigning the same, small a priori probability to a locus being associated with a trait. Based on this and the data, we can define the posterior probability of a locus being associated with a trait. One can then infer that a locus is associated with a trait if and only if the posterior probability of it being associated with that trait is at least 0.5. Assume that the number of tests is very large (e.g., we have data from a very large number of loci). Two types of sparsity are considered. Under extreme sparsity, the number of loci associated with a trait does not increase, even when the number of loci increases. Under standard sparsity, the number of loci associated with a trait increases at a lower rate than the total number of loci considered. Given a large number of tests under the assumption of extreme sparsity, inference based on the Bonferroni procedure is almost equivalent to inference based on Bayesian decision theory. Similarly, the BH procedure is almost equivalent to inference based on Bayesian decision theory under either extreme or standard sparsity. In addition, these testing procedures have the advantage of not needing (or having to infer) information regarding the proportion of loci that are actually associated with a trait. Hence, these testing procedures have very desirable properties in the statistical problems associated with GWAS, where a very large number of loci are considered and generally a very small proportion of loci are real factors.

In the case of model selection, the goal is not just to find which loci are associated with a given trait, but describe how those loci are associated with that trait. Again, when there are a large number of variables (loci), classical statistical methods (e.g., regression) tend to include more variables in the model than they should. Also, in many cases, regression methods may not even work, since in many problems from genetics, the number of loci is greater than the number of individuals. Any good model should possess the following two characteristics: (i) give an accurate description of the data, (ii) be relatively simple (parsimonious). Akaike's Information Criterion (AIC) and the Bayesian Information Criterion (BIC) are based on maximizing a function given by the log-likelihood of the data given the model (a measure of how the model fits the data) minus a penalty function based on the number of variables included in the model (a measure of the complexity of a model). The specific goals of these approaches are discussed. AIC is specifically designed to produce accurate predictions, while BIC is specifically designed to infer what variables are associated with the trait of interest. However, under the assumption of sparsity, BIC tends to include too many variables in the model. Hence, we consider adaptations of BIC to such scenarios. Section 3.2 also considers the LASSO, elastic net, and SLOPE methods of model selection. These approaches can be thought of as adaptations of AIC and BIC, since they can be defined in terms of maximizing a penalized likelihood function. However, under these three approaches, the penalty function depends on the magnitudes of effects of the variables included in the model, and not just on their number.

Chapter 4 concentrates on QTL mapping. Section 4.1 considers single marker tests, i.e., tests which choose between one of the following two hypotheses: (i) the null hypothesis that a locus is not associated with a quantitative trait and (ii) the alternative hypothesis that a locus is associated with a quantitative trait. Classically, in such tests we have information regarding the genotypes of n individuals at m markers. In general, a QTL will not be located at the same position as the marker. However, a strong association between the genotype of a marker and a quantitative trait indicates that it is very likely that a QTL is located close to that marker. As we are usually dealing with a number of markers, we should adopt a procedure to take multiple testing effects into account. For example, if we suspect that there is a single QTL on a particular chromosome, then we can apply the Bonferroni procedure to control FWER. The Bonferroni procedure makes no assumption regarding the correlation between test statistics. However, intuitively these test statistics will be naturally correlated, since if there exists an association between the genotype of a marker and the quantitative trait in question, then we should expect a similar association between the genotype of a neighboring marker and the quantitative trait. Hence, we also consider improvements to the standard Bonferroni procedure based on the correlation between the test statistics and compare the results from applying these procedures.

Section 4.2 considers more advanced methods of QTL mapping. The first is interval mapping, which estimates the position of a QTL that maximizes the likelihood of the data (the fit to the data) on the basis of an experimental population. Suppose a QTL is located between two markers. Given the genotypes at the markers, we can calculate the probabilities of the possible genotypes at the QTL based on the probability of crossover occurring. The distribution of the trait in the population can thus be interpreted as a mixture of conditional distributions given the genotype at this QTL. Using an iterative procedure, we can then calculate the likelihood of the data given that there is a QTL at a given position. Maximizing this likelihood function gives us an estimate of the location of a QTL. The second method is regression interval mapping. Using this approach, the genotype at each marker is coded numerically and standard regression can be used to test whether there is a QTL at the site of the marker. At sites that do not correspond to a marker, we can derive the expected value of such a numerical code for the genotype given the genotypes at the neighboring markers. In this case, the estimate of the position of the QTL corresponds to the largest realization of the test statistic for the presence of a QTL at a given position. This approach is much simpler to implement than interval mapping, but a comparison of the two approaches shows that they give very similar results.

Section 4.3 presents methods of model selection. The approaches described immediately above essentially assume that there is one QTL on a given chromosome. However, often we have data from different chromosomes, there can be several QTLs on a single chromosome and there may be interactions between various QTLs (i.e., the effect of a set of QTLs is not simply the sum of the individual effects). Hence, in practical situations the number of potential regressor variables will be very large (often larger than the number of individuals). Hence, in such problems, we should adapt procedures based on the adaptations of BIC considered in Sect. 3.2.

Section 4.4 shows that logic regression can apply the theory of logical expressions to express interactions in a simpler and more intuitive way than standard approaches based on linear regression. Although the number of possible models increases when such an approach is used, the increased power obtained using such an approach is sufficiently large to outweigh any possible losses from the need to control the false discovery rate.

Section 4.5 briefly describes how modifications of the Bayes information criterion can be applied in a Bayesian approach to statistical inference.

Section 5 presents Genome Wide Association Studies (GWAS). Such studies have come into prominence due to the data available from genome sequencers, which read the nucleotides making up an individual's DNA sequence. GWAS use the information from so-called single nucleotide polymorphisms (SNPs), which are positions in the sequence at which various nucleotides are observed within a single population. At such positions, in general, two variants are observed within a population. In this case, the genotype of an individual is given by the pair of variants observed. Denoting the two variants by a and A, the possible genotypes are aa, aA and AA. The processes involved in genome sequencing are stochastic; Sect. 5.1 presents the concepts behind inferring the genotype present at a given locus.

GWAS analyze the association between traits (which can be discrete or continuous) and the genotypes at SNPs. In general, the number of SNPs observed is much larger than both the number of individuals observed and the number of markers observed in QTL mapping. This implies that the problems inherent in multiple testing are much more apparent in GWAS. In Sect. 5.2, we consider single marker tests. Adopting such an approach, we carry out a series of m tests, where the null hypothesis in the i-th test is that the i-th SNP is not associated with the trait in question and the alternative is that the i-th SNP is associated with that trait. Various models of genetic association are considered. For example, it is possible that variant a dominates variant A, so that when considering the association between the genotype and a given trait, those of genotype aA do not differ on average from those of genotype aa, but do differ from those of genotype *AA*. Various tests of association are considered, including the standard χ^2 test of association which makes no assumptions regarding the form of any association and tests based on three different types of genetic association. In addition, we consider a single test based on a combination of these three tests.

Corrections for the effects of multiple testing are obviously of prime importance. Very often, GWAS are carried out in two stages. In the first stage, a very large number of SNPs from across the genome are considered. Testing is used to choose a set of SNPs for further investigation. Since such a two-step procedure is adopted, it is often sensible to use a more liberal correction procedure to avoid the loss of power which would result from adopting an essentially stricter procedure.

One problem associated with an approach based on single marker tests is that, in non-experimental populations, the observed frequency of the rare variant at an SNP may be very small. In such cases, the power of single marker tests to detect associations will be very small, especially when correction is made for the effects of multiple testing. One way of dealing with this problem is to group information from neighboring SNPs. We discuss possible ways of doing this and the problems involved with such an approach.

GWAS is often applied to non-experimental populations, in particular, human populations. However, such a population may have a structure, i.e., individuals are more likely to pair with those from the same subpopulation. Subpopulations may be based, e.g., on class and/or ethnicity. An approach to correcting test statistics due to population structure is described, together with a brief description of a method for analyzing population structure based on principal component analysis. Since phenotypes can depend on such factors as age and sex, we consider how such factors can be included into models describing a phenotype.

Since genes may interact in determining traits, carrying out single marker tests is a simplistic approach to GWAS and thus in Sect. 5.3 we consider model selection. In particular, the effects of individual genes may be relatively small and thus single marker tests will very often fail to detect a real effect. Hence, we consider more general models for a quantitative trait based on genetic (and possibly demographic) information. In such cases, the number of possible models is huge. Also, when there are a number of SNPs affecting a trait, the random associations between these SNPs and variants observed at other SNPs can often lead to an inflated false discovery rate, even when appropriate procedures are used. Three software packages for the analysis of GWAS are described are compared: HYPERLASSO, GWASelect, and MOSGWA.

Section 5.4 takes a slightly closer look at a situation where the population has a very specific structure. In recent times, many human populations which had been previously separated have become mixed. The genetic makeup of such populations is somewhat similar to that of experimentally produced crosses. Admixture mapping is an approach which uses this structure to search more effectively for SNPs associated with a particular trait.

Section 5.5 looks at the problem of detecting interaction between SNPs in their effect on a phenotype. Since considering the possibility of interaction greatly increases the number of possible models in the case of QTL mapping, in the case of GWAS, the number of possible models is simply huge. This section briefly considers application of the classical approach of analysis of variance and logic regression, also considered previously in Chap. 4, to the detection of such interactions. When a phenotype is dichotomous, one natural approach to detecting interactions between SNPs is to split combinations of genotypes into "high risk" and "low risk" categories, thus reducing the dimensionality of the problem. This approach is known as Multifactor Dimensionality Reduction (MDR). It should be noted that this approach can be adapted to the analysis of continuous phenotypes.

Section 5.6 gives a comparison of several methods for analyzing genetic effects on a dichotomous phenotype. Nearly all of the methods considered are adaptions of models considered in this book. However, the approach that has the greatest power to detect interactions, while still retaining reasonable power to detect individual effects, is different in its nature. It is specifically designed to analyze the joint distribution of a large number of discrete variables. However, the method used to select the appropriate model is very strongly embedded in the ideas that run through the whole book. In statistical genetics, just as in any other kind of research, cross fertilization of ideas is a key to scientific advance. This section ends with some brief thoughts on how GWAS will evolve in the near future.

References

1. Watson, J.D., Crick, F.H.: Molecular structure of nucleic acids. Nature **171**(4356), 737–738 (1953)
2. Kerem, B.S., Rommens, J.M., Buchanan, J.A., et al.: Identification of the cystic fibrosis gene: genetic analysis. Science **245**(4922), 1073–1080 (1989)
3. Sasiadek, M.M., Stembalska-Kozlowska, A., Smigiel, R., Ramsey, D., Kayademir, T., Blin, N.: Impairment of MLH1 and CDKN2A in oncogenesis of laryngeal cancer. Br. J. Cancer **90**(8), 1594–1599 (2004)
4. Schlade-Bartusiak, K., Rozik, K., Laczmanska, I., Ramsey, D., Sasiadek, M.: Influence of GSTT1, mEH, CYP2E1 and RAD51 polymorphisms on diepoxybutane-induced SCE frequency in cultured human lymphocytes. Mutat. Res. Genet. Toxicol. Environ. Mutagenesis **558**(1), 121–130 (2004)
5. Silventoinen, K., Kaprio, J., Lahelma, E., Koskenvuo, M.: Relative effect of genetic and environmental factors on body height: differences across birth cohorts among Finnish men and women. Am. J. Public Health **90**(4), 627

Chapter 2
A Primer in Genetics

2.1 Basic Biology

2.1.1 Phenotypes and Genotypes

A phenotype is any observable characteristic of an organism. Phenotypes of interest could be, for example, height, weight, blood pressure, blood type, eye color, disease status, the size of a plant's fruits, or the amount of milk given by a cow. Typically, one observes quite a large amount of variety in phenotypes between individuals of the same species. Phenotypes are influenced by both genetic and environmental factors. A great proportion of current biological research consists of trying to get a better understanding of the genetic factors involved.

In eukaryotes (organisms composed of cells with a nucleus and organelles), including plants, animals, or fungi, most of the genetic material is contained in the cell nucleus. This material is organized in **deoxyribonucleic acid (DNA)** structures called **chromosomes**. DNA consists of two long polymers of simple units called nucleotides. One element of a nucleotide is the so-called nucleobase (nitrogenous base). There are four primary DNA-bases: cytosine, guanine, adenine, and thymine, abbreviated as C, G, A, and T, respectively. Pairs of DNA strands are joined together by hydrogen bonds between complementary bases: A with T, and C with G. Therefore, the sequence of nucleotides in one strand can be determined by the sequence of nucleotides in the other (complementary) strand. The backbone of a DNA strand is made from phosphates and sugars joined by ester bonds between the third and fifth carbon atoms of adjacent sugar rings. The corresponding ends of DNA strands are called the $5'$ (five prime) and $3'$ (three prime) ends. Such a pair of DNA strands are orientated in opposite directions, $3'-5'$ and $5'-3'$. Therefore, they are called **antiparallel**.

A pair of DNA strands form a structure known as the **double helix**, illustrated in Fig. 2.1. However, for the purpose of many statistical and bioinformatical analyses, chromosomes are simply represented as sequences, where each element is the letter

© Springer-Verlag London 2016
F. Frommlet et al., *Phenotypes and Genotypes*, Computational Biology 18,
DOI 10.1007/978-1-4471-5310-8_2

Fig. 2.1 An illustration of
the double helix structure
and two antiparallel
sequences of nucleobases

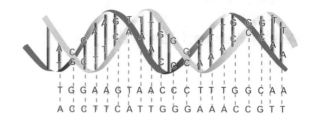

$$T\ G\ G\ A\ A\ G\ T\ A\ A\ C\ C\ C\ T\ T\ T\ G\ G\ C\ A\ A$$
$$A\ C\ C\ T\ T\ C\ A\ T\ T\ G\ G\ G\ A\ A\ A\ C\ C\ G\ T\ T$$

corresponding to the nuclear base (C, G, A, or T) at the corresponding position in
one of the strands.

In the process of **transcription**, some sections of DNA, called **genes**, are tran-
scribed into complementary copies of ribonucleic acid (RNA). Since RNA is single
stranded, only one strand of DNA is used in the transcription process. The resulting
RNA strand is complementary and antiparallel to the "parental" DNA strand, with
thymine (T) being replaced by uracil (U). As a result, the RNA sequence is identical
(except for T being replaced by U) to the complementary sequence of the parental
DNA strand.

If a gene encodes a protein, then the resulting messenger RNA (mRNA) is used
to create that protein through the process of **translation**. Proteins can be viewed
as chains of amino acids, where certain triplets of mRNA are translated into specific amino acids. In eukaryotes there is a further modification of RNA between the
processes of transcription and translation, which is called **splicing**. Here, parts of the
RNA, so-called **introns**, are removed and the remaining parts, called **exons**, become
attached to each other. After splicing, the mRNA consists of a sequence of triplets,
which directly translate into the amino acids forming the protein expressed by the
gene in question.

The DNA sequence corresponding to an mRNA sequence is called a **sense** strand.
Thus, as explained above, a sense DNA sequence is complementary to the corresponding parental (**antisense**) DNA sequence. Both strands of DNA can contain
sense and antisense sequences. Antisense RNA sequences are also produced, but
their function is not yet well known. Proteins, as well as functional RNA chains,
created via transcription and translation play an important role in biological systems
and influence many phenotypes.

The process of gene expression depends not only on the coding region, but also
on the regulatory sequences that direct and regulate the synthesis of gene products.
Cis-regulatory sequences are located in the close vicinity of the corresponding
gene. They are typically binding sites for transcription factors (usually proteins),
which regulate gene expression. **Trans-regulatory** elements are DNA sequences
that encode these **transcription factors** and are not necessarily close to the gene in
question. They may even be found on different chromosomes.

The DNA sequences of different individuals from a given species are almost identical. For example, in humans 99.9 % of all DNA-bases match. However, there still
exist a large number of **polymorphic** loci, at which differences between individuals from a given species can be observed. The variants observed at such a locus

are called **alleles**, where the most prominent examples of such genetic variation are **single nucleotide polymorphisms** (SNPs) and **copy number variations** (CNVs). SNPs refer to specific positions in a chromosome where different nucleobases are observed, the result of a so-called point mutation. Copy number variation refers to relatively long stretches of DNA which are repeated a different number of times in various individuals. In particular, insertions, deletions, and duplications of DNA stretches are classified as CNVs. If the DNA section corresponding to a CNV includes a gene, it will result in different gene expression patterns. **Microsatellites** are also classical examples of genetic polymorphisms, where very short DNA patterns are repeated a number of times, and the number of repetitions varies between individuals.

The number of **homologous** chromosomes, which at a given locus contain genes corresponding to the same characteristic, varies between different species. **Haploid** organisms, such as male bees, wasps, and ants, have just one set of chromosomes (i.e., just one copy of each gene). The majority of all animals, including humans, are **diploid**, i.e., they have two sets of chromosomes, one set inherited from each parent. In diploid organisms an individual's **genotype** at a given locus is defined by the pair of alleles residing at this locus on the two homologous chromosomes. For example, consider a biallelic locus with alleles A and a. Then there exist three possible genotypes: AA, Aa, and aa. An individual carrying two identical alleles at a given locus is called **homozygous** at this locus, whereas an individual with two different alleles is **heterozygous**. There also exist many organisms which are **polyploid**, meaning that they have more than two homologous chromosomes. Polyploid organisms are common among plants, e.g., the potato, cabbage, strawberry, and apple. In this book, we will mainly focus on methods for localizing genes in diploid organisms.

A **haplotype** is an ordered sequence of nucleobases appearing on the same chromosome. For example, a haploid organism inherits a maternal haplotype and a paternal haplotype, which together define the genotypes at the corresponding loci. When an individual is genotyped, generally we do not know which parent each allele came from. In this case, we say that the genotypes are unphased. Hence, it might be necessary to infer the haplotypes from the genotype data (in other words, determine the phase). One of the most popular algorithms for phasing is FASTPHASE [113], which applies maximum likelihood methods to predict haplotypes. In this book, we will mainly focus on statistical methods which make use of genotype data, although many of the statistical methods described in Chap. 5 can be extended to phased haplotype data. For illustrative purposes, Table 2.1 gives a simple example of unphased genotypes at 10 markers, and two phased haplotypes corresponding to these genotypes.

Table 2.1 Unphased genotypes and phased haplotypes for 10 markers

Unphased	aA	BB	cC	dD	ee	ff	gG	hH	iI	JJ
From father	A	B	c	d	e	f	G	H	i	J
From mother	a	B	C	D	e	f	g	H	I	J

The genetic information defining gender is typically contained in **sex chromosomes**. Among diploid organisms, the XX/XY sex-determination system is the most common. In this system, females have two sex chromosomes of the same kind (XX), while males have two distinct sex chromosomes (XY). The X and Y sex chromosomes are different in size and shape from each other. The Y chromosome contains a gene called SRY (the sex determining region of Y) that determines maleness and can only be inherited from the father. In humans the X chromosome spans more than 153 million base pairs coding for approximately 2000 genes, while the Y chromosome spans about 58 million base pairs and contains 86 genes, which code for only 23 distinct proteins. Traits that are inherited via the Y chromosome are called holandric traits. The remaining chromosomes, i.e., those not related to gender, are called **autosomes**.

2.1.2   Meiosis and Crossover

In many organisms genetic material is passed on from parents to offspring through the process of **sexual** reproduction. **Meiosis** is the biological process via which gametes or spores are produced. In diploid organisms meiosis transforms one diploid cell into four haploid gametes. Before meiosis, the paternal and maternal copies of each chromosome duplicate themselves and form two pairs of identical **sister chromatids**, where each pair are joined together at the centromere. Then the maternal and paternal homologues pair with each other. Occasionally, genetic material is exchanged between a paternal and maternal chromosome in events called **crossover**, where matching regions of both chromosomes break and then reconnect with the other chromosome. In the final stage of division, the cell is divided into four gametes, each containing one homologous chromatid. After crossover, such a gamete may contain genetic material from both maternal and paternal homologues. If the genetic material at two loci comes from different parents, then we say that **recombination** has occurred between these two loci. If two loci reside on different chromosomes, then the probability of recombination between them is equal to 1/2. If two loci reside on the same chromosome, then recombination results from an odd number of crossovers. Recombination increases the genetic diversity of a population. An illustration of meiosis is given in Fig. 2.2.

2.1.3   Genetic Distance

The physical distance between two loci is often expressed in terms of the number of nucleotide bases lying between them. However, for the purpose of gene mapping, scientists normally use genetic maps, which express distance in terms of probabilistic units. Genetic distance is often measured in **Morgans**. A distance of 1 Morgan means that the expected number of crossovers between two loci in a single meiosis

Fig. 2.2 A graphical
illustration of meiosis

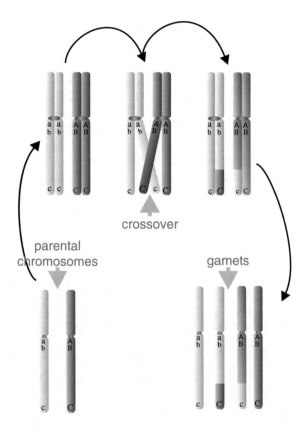

is equal to 1. There exist several **mapping functions**, which relate the probability of
recombination to the genetic distance.

2.1.4 The Haldane Mapping Function

In practice, the **Haldane function** is the most frequently used mapping function. It
assumes a **lack of interference**, which means that crossover events occur independently of each other. Under this assumption, the number of crossovers on a given
piece of a chromosome can be modeled by the Poisson distribution (see Sects. 6.3.1
and 6.5.2).

Let the distance between two loci be equal to d Morgans. Then the number X of
crossovers between these loci has a Poisson distribution with mean d,

$$P(X = k) = \frac{d^k}{k!} \exp(-d), \quad \text{for} \quad k \in \{0, 1, 2, \ldots\}.$$

Thus the probability of recombination between these two loci, r, is given by the formula

$$r = \sum_{k=0}^{\infty} P(X = 2k + 1) = \sum_{k=0}^{\infty} \frac{d^{2k+1} \exp(-d)}{(2k + 1)!}$$

$$= \sinh(d) \, \exp(-d) = \frac{e^d - e^{-d}}{2} \, e^{-d} = \frac{1}{2}(1 - e^{-2d}). \qquad (2.1)$$

Note that according to this formula $0 \le r \le 1/2$, where $r = 0$ corresponds to a genetic distance of 0, and $r = 1/2$ to an infinite genetic distance. As the genetic distance increases, the recombination rate converges rapidly to 1/2, i.e., for large d the recombination rate is very close to 1/2. For loci on different chromosomes, one usually defines $r = 1/2$. When $r < 1/2$, then we say that two loci are linked (i.e., lie on the same chromosome).

Solving Eq. 2.1 for d yields the

Haldane mapping function:

Let r be the probability of recombination between two loci and d be the genetic distance between these loci (in M). Under the assumption of no interference, it follows that $d = H(r)$, where

$$H(r) := -(1/2) \ln(1 - 2r) \qquad (2.2)$$

is the Haldane mapping function.

2.1.5   Interference and Other Mapping Functions

Consider three genetic loci at positions $L_1 < L_2 < L_3$ on the same chromosome, and denote by R_{ij} the event that recombination occurs between loci i and j. Then it is immediately clear that

$$R_{13} = (R_{12} \cup R_{23}) \backslash (R_{12} \cap R_{23}).$$

If the events R_{12} and R_{23} are independent, then the probabilities r_{ij} of recombination between loci i and j satisfy

$$r_{13} = r_{12} + r_{23} - 2r_{12}r_{23}.$$

This equality is fundamental to defining the Poisson process underlying the derivation of the Haldane mapping function. However, in practice, one often observes that the occurrence of crossover at one locus reduces the chance of crossover at neighboring loci. Due to this kind of **interference**, the probability of a double crossover in a given interval is smaller than that estimated by applying the Poisson process and, consequently, the probability of recombination between neighboring loci is larger than the estimate provided by the Haldane function. In general, the recombination fractions will satisfy equation

$$r_{13} = r_{12} + r_{23} - 2Cr_{12}r_{23},$$

where $C \in [0, 1]$ is the coefficient of **coincidence** and $I = 1 - C$ is the coefficient of interference. In the case $I = 0$, there is no interference, resulting in the genetic distance being described by the Haldane function (2.2). The other extreme situation, $I = 1$, corresponds to complete interference, which eliminates the possibility of more than one crossover on any chromosome. In this case, the genetic distance d and the probability of recombination r are connected via the simple

Morgan mapping function:

In the case of complete interference, one has

$$d = M(r) := r.$$

Note that according to the assumption of complete interference, the expected number of crossovers on each chromosome cannot exceed 1, which limits the maximal length of a chromosome to 1 Morgan.

Another popular mapping function assumes $I = 0.5$. This intermediate choice of the interference parameter gives rise to the

Kosambi mapping function:

$$d = K(r) = \frac{1}{4} \ln \left[\frac{1 + 2r}{1 - 2r} \right].$$

As in the case of the Haldane mapping function, the recombination fraction $1/2$ corresponds to an infinitely long chromosome. However, for any given genetic distance, the recombination fraction according to the Kosambi function exceeds the recombination fraction given by the Haldane function.

The relationship between the recombination fraction and the genetic distance according to the three mapping functions discussed above is presented in Fig. 2.3.

Fig. 2.3 The relationship
between the recombination
fraction and the genetic
distance according to
different mapping functions

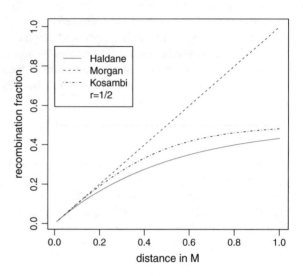

2.1.6 Markers and Genetic Maps

In Sect. 2.1.1 we briefly mentioned several possible types of genetic variation between
individuals of the same species, such as SNPs or CNVs. Such polymorphisms are of
primary importance in a large number of genetic studies. A polymorphic piece of a
DNA sequence with known location on a chromosome serves as a **genetic marker**.
Historically, the genotype at the earliest known genetic markers was determined by
observing the corresponding phenotypic trait, e.g., blood group or the color of flowers.
Today, a vast number of genetic markers are available, due to modern sequencing
techniques. Markers are used to construct genetic maps and as reference points, to
decide which parts of the genome have some influence on a given phenotypic trait.

There exist many experimental techniques for genotyping markers. Among the
most popular classical systems of markers, one should mention the following:
restriction fragment length polymorphisms (RFLPs), randomly amplified polymorphic DNA (RAPDs), and short tandem repeats (STRs), which are also known as
microsatellites. The alleles at microsatellite markers differ in the number of repli-
cations of short (1–6 base pairs) sequences of DNA. In comparison to RFLPs and
RAPDs, microsatellites have a substantially greater number of alleles and are more
useful for locating genes in natural **outbred** populations. Due to the development
of new DNA sequencing techniques, the determination of Copy Number Variations
(CNVs) has become possible, an approach which has gained large popularity in
recent years, especially in the context of human genetics.

The most popular system of markers in Genome Wide Association Studies
(GWAS) is based on single nucleotide polymorphisms (SNPs). A single nucleotide
polymorphism occurs when different nucleotides appear at a given single base pair
within a population. As an example, consider the following two sequenced DNA

fragments from different individuals, AGCCT and AGCTT. There is a SNP at the fourth position with alleles C and T. At the vast majority of SNPs, only two alleles are possible, therefore this system of markers is essentially considered to be biallelic. The allele which is more frequent in a population is called the **reference** (or **major**) allele, while the less frequent allele is called the **variant** (or **minor**) allele. SNPs have gained large popularity in recent years, due to the development of SNP microarrays. This technology facilitates the quick and relatively cheap genotyping of several hundred thousand SNPs at the same time. Often these microarrays include a comparable number of CNVs.

2.2 Types of Study

Depending on the organism under investigation and the trait of interest, there exist a variety of experimental designs and methods to locate influential genes. When dealing with organisms which reproduce quickly and can be experimentally crossed, one can use techniques which have been specifically developed for such **experimental populations**. Based on such crosses, statistical methods for detecting a qualitative trait locus (QTL—a gene influencing a quantitative trait) are usually referred to as **QTL mapping**. It is not practical to generate experimental populations of certain species, which is particularly true for the human species, for obvious reasons. In this case, the types of studies used fall into one of the two categories: **linkage analysis**, which is based on data from families (pedigrees), and **association studies**, which are quite often performed with so-called outbred populations, which means that there are no close relatives within the study sample. The logic underlying association studies is rather similar to QTL mapping, whereas linkage analysis based on pedigree data is fundamentally different and will not be discussed in depth in this book. A good basic introduction to linkage analysis is given by [4].

2.2.1 Crossing Experiments

The starting point for all experimental populations are so-called **inbred lines**. These are obtained by mating only individuals from the same line over successive generations. Due to the fundamental laws of inheritance in small populations, which were described by Fisher [5], those alleles which are less frequent tend to get eliminated and after a number of generations the individuals from a given inbred line are genetically identical and homozygous at almost all loci (which means they have the same alleles at both chromosomes). These lines are typically chosen in such a way that one observes a large difference in both the phenotype of interest and the traits related to the markers between the two lines. Experimental populations are then obtained by crossing individuals from two distinct inbred lines.

Let us generically denote alleles from the first line by a, and alleles from the second line by A. The **F1 population** results from crossing individuals from both lines with each other. This generation is again genetically homogeneous, because each F1 individual has genotype Aa at each locus, which means that it has alleles from both inbred lines. In other words, each F1 individual is heterozygous at each locus. Further crosses involving F1 individuals lead to the experimental populations used in genetic studies. Depending on the exact strategy followed, one obtains different experimental designs, among which the most popular are the **backcross** design, the **intercross** design, and **recombinant inbred lines**.

2.2.1.1   Backcross Design

Using a backcross design, individuals from the F1 population are crossed with individuals from one of their parental inbred lines (line **P1** in Fig. 2.4). The resulting individuals form a backcross population (**BC1** in Fig. 2.4). At each genetic locus there are only two possible genotypes, either homozygous (with both alleles from the parental inbred line P1) or heterozygous. Thus, at each locus the genotype of BC1 individuals is fully determined by the allele inherited from its F1 parent.

It is often convenient to encode the genotype at a genetic locus by $X = 0$ in the case of homozygosity, and by $X = 1$ otherwise. For theoretical considerations, the state of the genotype can be thus interpreted as a random variable. Due to the crossing strategy of the backcross design, it follows that $P(X = 0) = P(X = 1) = 1/2$. Thus X is **Bernoulli distributed** with "success probability" $p = 1/2$. It immediately follows that $E(X) = 1/2$ and $\text{Var}(X) = 1/4$.

Now let us consider two genetic loci, L_1 and L_2, residing on the same chromosome. It turns out that if L_1 and L_2 are close to each other, then it is likely that an individual from the BC1 generation will be either homozygous at both loci or heterozygous at both loci. In fact, the conditional probability that an individual is homozygous at L_1 given that it is heterozygous at L_2 is simply the probability of recombination between these two loci. In mathematical terms, this can be expressed as

$$P(X_1 = 0 | X_2 = 1) = r, \quad P(X_1 = 1 | X_2 = 1) = 1 - r,$$

Fig. 2.4 Descriptions of the backcross (*left*) and intercross (*right*) designs

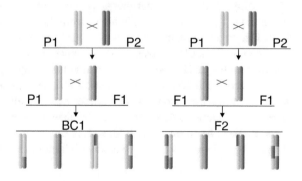

where X_1 and X_2 denote the genotypes at L_1 and L_2, respectively. The genetic distance between loci then directly translates into the correlation between their genotypes according to the formula

$$\text{Corr}(X_1, X_2) = 1 - 2r. \tag{2.3}$$

This is easily obtained by computing the covariance between the genotypes

$$
\begin{aligned}
\text{Cov}(X_1, X_2) &= E(X_1 X_2) - E(X_1)E(X_2) \\
&= P(X_1 = 1, X_2 = 1) - 1/4 \\
&= P(X_1 = 1 | X_2 = 1)P(X_2 = 1) - 1/4 = (1-r)/2 - 1/4 = \frac{(1-2r)}{4}.
\end{aligned}
$$

2.2.1.2 Intercross Design

Using the intercross design, often denoted as **F2**, individuals from the F1 population are crossed with each other (see Fig. 2.4). Individuals from the F2 population can have any of the three possible genotypes at each locus: AA, Aa or aa. As in the case of a backcross design, one can easily calculate the conditional probabilities of a certain genotype at locus L_2 given the genotype at locus L_1 as a function of the distance between these two loci. These probabilities are given in Table 2.2.

Using the intercross design, the classical coding of genotypes is defined by the Cockerham model, which uses two state variables X and Z for the genotype at each locus (see Table 2.3). It will be shown in Sect. 2.2.2 that the variable X is

Table 2.2 Intercross (or **F2**) design: conditional probabilities of genotypes at locus L_2 given the genotype at L_1 and the probability r of recombination between these two loci

L_1	L_2 AA	Aa	aa
AA	$(1-r)^2$	$2r(1-r)$	r^2
Aa	$r(1-r)$	$r^2 + (1-r)^2$	$r(1-r)$
aa	r^2	$2r(1-r)$	$(1-r)^2$

Table 2.3 Coding of genotypes using the Cockerham model

Genotype	Dummy X	Variables Z
AA	-1	-0.5
Aa	0	0.5
aa	1	-0.5

Fig. 2.5 The correlation coefficient between variables coding for additive and dominance effects as a function of the genetic distance

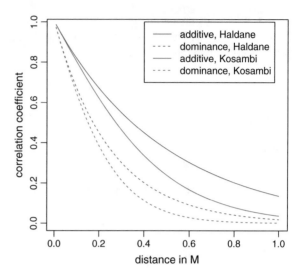

associated with the additive effects of a QTL (a measure of the mean difference in the value of the trait between the two types of homozygote). Similarly, the variable Z is associated with the dominance effect (this is zero if the mean value of the trait among heterozygotes is exactly half way between the mean values in the homozygote groups, i.e., neither allele dominates the other). Based on the conditional probabilities from Table 2.2, it is a simple exercise to compute the correlation between these two dummy variables according to the genetic distance:

$$\text{Corr}(X_1, X_2) = 1 - 2r, \quad \text{Corr}(Z_1, Z_2) = 4(r - 0.5)^2, \tag{2.4}$$

$$\text{Corr}(X_1, Z_1) = \text{Corr}(X_1, Z_2) = 0,$$

where r represents the probability of recombination between the two loci.

The correlation coefficient between these state variables as a function of the genetic distance is illustrated in Fig. 2.5. It can be observed that the correlation coefficient between variables corresponding to dominance effects decays faster than the correlation coefficient between variables describing additive effects. Also, these correlation coefficients decay faster for the Kosambi mapping function than the Haldane mapping function.

2.2.1.3   Recombinant Inbred Lines

Recombinant inbred lines are obtained by the multiple crossing of close relatives or by the multiple self-fertilizing of individuals from an F2 population. Due to the elimination of "rare" alleles, recombinant inbred lines consist of individuals who are homozygous at almost every locus, but can contain alleles from different parental

Table 2.4 Recombinant inbred lines: Probabilities R of obtaining two distinct genotypes at loci L_1 and L_2, given the probability r of recombination between these loci

Type of inbred line	$P(X_1 \neq X_2)$
Self-fertilizing	$2r/(1 + 2r)$
X chromosome, sibling mating	$(8/3)r/(1 + 4r)$
Autosomes, sibling mating	$4r/(1 + 6r)$

inbred lines at different loci. Because only two genotypic states are possible at each locus, the methods used for the statistical analysis of recombinant inbred lines is, to a certain extent, similar to the analysis of a backcross design. However, the correlation between genetic loci is different for recombinant inbred lines. Let r be the recombination fraction between loci L_1 and L_2. According to [7] (see also [1]), the probabilities of obtaining two distinct genotypes X_1 and X_2 at these locations are given in Table 2.4. For the backcross design, the probability of observing two different genotypes is simply given by r, which is smaller than for any recombinant inbred line. As a result, the backcross design needs less markers per chromosome to detect a QTL, but it also gives less precision with respect to the exact location.

Traditionally, recombinant inbred lines are generated from only two parental generations. In a more recent project, recombinant inbred lines were derived from a genetically diverse set of eight founder inbred mouse lines [2, 9]. This experimental population was particularly designed to mimic the genetic diversity of humans, while keeping the advantages of a controlled population.

2.2.2 The Basics of QTL Mapping

A **quantitative trait locus (QTL)** is a location on the genome which hosts a gene that influences a certain quantitative trait. The major goal of QTL mapping is to identify such regions by means of statistical analysis. Considering an individual from a population, its trait value Y is a random variable which depends on the genetic background, as well as on many environmental factors. The **broad heritability** of a trait is defined to be the proportion of the trait's variance which can be explained by genetic factors.

To explain the basic principles of QTL mapping, we start with an extremely simple scenario. Consider a backcross design, and assume that there is exactly one QTL which influences the trait. Let $\mu_1 = E(Y|QTL = AA)$ denote the mean value of the trait when the QTL genotype is AA. Analogously, let $\mu_2 = E(Y|QTL = Aa)$. The coefficient $\beta = \mu_2 - \mu_1$ is called the effect size of the QTL. This is simply the expected increase in the trait value when allele a is substituted by A. Now consider a marker M which lies on the same chromosome as the QTL at a distance such that the probability of recombination between M and the QTL equals r. According to the

law of total probability, the expected value of the trait given the marker genotype is
given by

$$E(Y|M = AA) = \mu_1(1 - r) + \mu_2 r \, , \qquad (2.5)$$

$$E(Y|M = Aa) = \mu_1 r + \mu_2(1 - r) \, . \qquad (2.6)$$

Thus

$$E(Y|M = Aa) - E(Y|M = AA) = (\mu_2 - \mu_1)(1 - 2r)$$

and the difference between the mean trait values for individuals with different marker
genotypes is different from zero as long as $r < 1/2$ (which means that the marker
and the QTL are linked). Clearly, conditional on the marker genotype, the effect size
is larger when the marker is closer to the QTL. This enables the detection of a QTL
by identifying markers whose genotypes are associated with the trait. The details of
statistical tests which can be used for this purpose will be discussed in Chap. 4.

As a second example, consider an intercross population with exactly one QTL.
We define

$$\mu_1 = E(Y|QTL = AA), \quad \mu_2 = E(Y|QTL = Aa), \quad \text{and } \mu_3 = E(Y|QTL = aa).$$

The QTL is said to have a purely **additive effect** if $\mu_2 = (\mu_1 + \mu_3)/2$, which means
that the average difference in the values of traits between individuals with genotypes
AA and those with Aa is exactly the same as the average difference in the values of traits between individuals with genotypes Aa and those with aa. Otherwise,
the QTL is said to have a **dominance effect**, defined by $\gamma = \mu_2 - (\mu_1 + \mu_3)/2$.
If $\mu_1 \neq \mu_2 = \mu_3$, then the allele a is said to be dominant with respect to A. On
the other hand, if $\mu_1 = \mu_2 \neq \mu_3$, then the allele a is called recessive. In general,
the additive effect of a QTL is defined by $\beta = (\mu_3 - \mu_1)/2$. This corresponds to the
coefficient of X in a regression model for the value of the trait and γ corresponds to
the coefficient of Z. A graphical representation of additive and dominance effects is
presented in Fig. 2.6. It is also possible that $\mu_2 \geq \max\{\mu_1, \mu_3\}$ or $\mu_2 \leq \min\{\mu_1, \mu_3\}$,
which is referred to as *overdominance*.

In experimental populations there exists a very strong association between neighboring loci, as can be seen, for example, in Fig. 2.5. Therefore, to detect causal genes,
it is usually enough to use approximately 10 markers on each chromosome. Based
on the association studies discussed in the next sections, this enables us to minimize the problems resulting from multiple testing (see Sect. 3.1) and increase the
power to detect QTLs. However, the strong association between neighboring markers also results in the rather low precision of estimators of the location of QTLs in
experimental lines.

When compared to association studies, it is important to note that experimental
populations give a researcher control over the genetic composition of the population.
Usually, it is also much easier to control environmental influences on experimental
populations than it is in natural populations, which is crucial for association studies.

Fig. 2.6 Graphical
representation of the additive
effect β and the dominance
effect γ

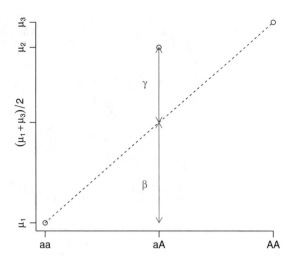

2.2.3 Association Studies

The most important insight gained from the previous section is that a QTL can be
detected if it is strongly linked to a genetic marker. This is a fundamental feature,
which enables us to detect QTLs in experimental populations, where the experimental
design gives us precise knowledge regarding the correlation between genetic loci on
the same chromosome. The general idea of association studies is very similar: genetic
markers strongly linked to influential genetic loci are expected to also be associated
with the phenotype in question.

However, in contrast to QTL mapping, association studies are often performed
with outbred populations, which means that the study sample does not include any
close relatives. For outbred populations, the correlation structure between genetic loci
is much more complicated than for experimental populations. There are no simple
formulas corresponding to Eq. (2.3) which express the correlation between genetic
loci as a simple function of the genetic distance. Instead, one analyzes **linkage
disequilibrium** (LD), which is a measure of the nonrandom association between
different genetic loci. For an overview of measures of LD see [3].

Two markers in LD cannot be treated as independent variables, since they are
correlated. The somewhat complex theory underlying LD in outbred populations
is a subject of the theory of population genetics. The most relevant source of LD
in association studies is the genetic linkage between markers located close to each
other on the same chromosome. As in experimental populations, a genetic marker
can be associated with a trait, although it does not directly affect that trait. It is
sufficient that such a marker is closely linked to a QTL. However, the local structure
of LD tends to be rather complicated, as can be seen in Fig. 2.7, which is a heatmap
based on data from the HapMap project. The color of this heatmap represents the
LD measure R^2, which is simply defined to be the square of the Pearson correlation

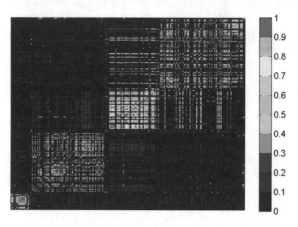

Fig. 2.7 Heatmap illustrating the LD pattern for 90 individuals from the CEU HapMap population (of Central European ancestry). The color code corresponds to the LD measure R^2 between the first 250 adjacent SNPs of the ENCODE region ENm010 (after removing duplicate sequences)

coefficient (see Sect. 7.4) between the genotypes of SNPs (coded as $X \in \{-1, 0, 1\}$ as in Sect. 2.2.1.2).

The HapMap project has been crucial in developing a better understanding of LD patterns within the human genome [13, 14]. In the second phase of the project, genotypes were obtained from 269 individuals from four populations: 90 Yoruba from Ibadan, Nigeria (YRI); 90 U.S. residents of Central European ancestry (CEU); 45 Japanese from Tokyo (JPT); and 44 Chinese from Beijing (CHB). The aim was to obtain a comprehensive map of the human genome. After completing phase II of the project in 2007, over 3.1 million SNPs had been genotyped. The ENCODE project [12] considered ten regions in particular, for which we have highly accurate genetic maps. Within these regions of approximately 500 kilo base pairs, almost all the SNPs known at the time were genotyped within this study group.

Other popular measures of pairwise LD in the population genetics literature are often based not on genotype information, but require additional knowledge about **phased haplotypes**, i.e., one has to know which copy of the parental chromosome each marker allele belongs to.

Let us consider two genetic loci, A and B, and the corresponding dummy variables Y_{Ai} and Y_{Bi}, which are equal to 1 if the allele at the ith phased haplotype at the corresponding locus is the reference one and 0 otherwise (e.g., $i = 1$ may correspond to the maternal haplotype and $i = 2$ to the paternal haplotype). Then the frequencies of the reference alleles at these loci are given by $p_A = \bar{Y}_A$ and $p_B = \bar{Y}_B$, where $\bar{Y}_A$ and $\bar{Y}_B$ denote the average of Y_{Ai} and Y_{Bi} over all $2n$ phased haplotypes (where n is the number of diploid individuals). Moreover, the percentage of haplotypes for which the reference allele appears at both locations is given by the average of the products of Y_{Ai} and Y_{Bi}; $p_{AB} = \overline{Y_A Y_B}$. Two classical measures of LD between A and B, Lewontine's

D' and r^2, are based on the statistic $D = Cov(Y_A, Y_B) = p_A p_B - p_{AB}$. According to Lewontine's D', D is scaled according to

$$D_{\max} = \begin{cases} (1 - p_A)(1 - p_B) & \text{if } D < 0 \\ \min\{p_A(1 - p_B), p_B(1 - p_A)\} & \text{if } D > 0, \end{cases} \tag{2.7}$$

i.e., $D' = D/D_{\max}$. To derive r^2, the square of D is divided by the product of the variances of Y_A and Y_B; $r^2 = \frac{D^2}{p_A(1-p_A)p_B(1-p_B)}$. Thus r^2 is simply the square of the standard Pearson correlation coefficient between Y_A and Y_B. Now, observe that the variable X_A describing the genotype at locus A can be represented as $X_A = Y_{A1} + Y_{A2} - 1$, where Y_{A1} and Y_{A2} are the dummy variables corresponding to the haplotypes. It is easy to check that when Y_{A1} and Y_{A2} are independent, then the population correlation coefficient between X_A and X_B equals the correlation coefficient between Y_A and Y_B. Therefore, in real life situations the correlation coefficient R^2 between the genotypes at two loci is typically very close to r^2.

Figure 2.7 nicely illustrates the rather complex local LD patterns in humans. There appears to be some kind of block structure, where a number of neighboring SNPs tend to be all highly correlated with each other. However, this pattern of correlation is not necessarily completely in accordance with physical distance, which has already been pointed out in [13]. For example, the large LD block in the middle is pervaded by several blue stripes. This indicates that there is a set of highly correlated SNPs, but in between there is a number of other SNPs which are not at all correlated with that set.

Such models of haplotype blocks provide a useful first approach to looking at LD structure in outbred populations. However, in reality the situation appears to be even more complicated [14]. In any case, the mere existence of local LD enables us to adapt the idea of association studies to outbred populations. Compared with the experimental populations discussed previously, linkage is only observed at a much smaller scale of genetic distance. As a result, a much larger number of genetic markers are necessary to perform association studies. On the other hand, association studies enable much greater precision in localizing influential genes, precisely because LD does not extend over larger regions of the genome.

Association studies have a long history within genetic research, although linkage studies based on family pedigrees used to be more prominent in the past. While linkage studies have been quite successful for locating the QTLs of traits that are controlled by a single locus (Mendelian traits), they have turned out to be far less successful for locating QTLs for traits that are determined in a more complex manner. In the twentieth century, association studies were often rather limited by the fact that an insufficient number of genetic markers were available. In many cases, only a relatively small number of candidate polymorphisms, which were suspected in advance to be related to some trait, could be analyzed. In **candidate gene studies**, researchers are only interested in polymorphisms lying within the region of genes suspected to affect the trait in question. Preselection of the candidate genes is often based on a biological understanding of their functioning. On the other hand, there might exist some knowledge stemming from previous studies about genetic regions associated with a trait. In such cases, a follow up association study might run under the name of "fine mapping."

Today, a sufficiently dense map of genetic markers (usually SNPs) is available to carry out **genome wide association studies (GWAS)** among many species. In particular, we have already mentioned the HapMap project, based on the human population, which has mapped several million SNPs across the whole genome. More recently, the number of known SNPs has been further increased by the 1000 genomes project [17], which aims at sequencing the whole genome of 2500 individuals from about 25 populations around the world. Knowledge about such polymorphisms is important when designing a genetic map, but equally important is the question of how to determine the genetic variants of individuals participating in a study. Within the last two decades, microarray and sequencing technology have made rapid progress and brought down the costs of determining individuals' genotypes. In Chap. 5 we will discuss the technological aspects underlying GWAS in more detail, in particular, the microarray technology which enables determining millions of genetic variants in one experiment.

An interesting question in GWAS is whether it is really necessary to work with all known polymorphisms. Although the most recent technology allows us to genotype more than 2 million SNPs, this is still less than 20 % of the variants available today. On the other hand, it is well known that in regions of strong LD, a small number of phased haplotypes (so-called common haplotypes) comprise a very large majority of all haplotypes (see for example [19]). When designing an association study, one has to decide which SNPs one should genotype in patients. Similarly, when designing a SNP array, one has to decide which SNPs to put on that array. Clearly, it is not advisable to consider SNPs which are highly correlated, as they will each provide almost the same information. This leads to the idea of **tag SNP selection**: starting from an extensive set of SNPs, one looks for a minimal subset of SNPs, so-called tag SNPs, which contain as much information as possible. A large number of algorithms for tag SNP selection are available (see [6] for a brief review). A set of tag SNPs covering the whole genome can then be used to create SNP arrays.

In view of the latest developments in next generation sequencing, we can look ahead toward association studies where the complete genetic information regarding individuals is available. In theory, the question of LD would then be resolved, because all genetic variation (at the level of DNA) would be known. However, apart from the fact that occasional errors in sequencing lead to imperfect information, many other difficulties presently affecting GWAS will remain. In particular, the problem of multiple testing will become even worse. In Chap. 3, the statistical theory of multiple testing will be comprehensively discussed. Here, we only want to mention that due to the tremendous number of SNPs (or other types of genetic variation) considered in association studies, it is very likely that one observes an association between a trait and a marker just by chance. The only remedy against this intrinsic statistical problem is to perform very large scale studies, and it has become more or less standard in GWAS to consider study groups with several thousand participants.

2.2.3.1   Design Questions in Association Studies

So far, we have been mainly concerned with the basic ideas underlying population association studies. Ideally, these are based on a large number of unrelated individuals. Unrelated essentially means that relationships are distant enough so that no linkage due to the relatedness of individuals is observed. To guarantee that this is the case, it is advisable to perform statistical tests which rule out unknown family relationships between each pair of participants. Otherwise, the theoretical properties of the statistical tests applied might become distorted. For example, undetected relationships between participants of a study could increase the number of false positives.

Association studies can be performed based on quantitative traits (as in QTL mapping), but more often one deals with dichotomous traits, usually characterizing an individual's status with respect to a certain disease. In so-called **case control studies**, one considers samples of affected cases and unaffected controls. Often, it is relatively easy to recruit cases for association studies, but it might be more difficult to find controls who are prepared to be genotyped. Also, in view of financial restrictions, it is often easier to use a control group from the general population, for whom genomic data should be available in a reference database. However, using such a **case random design**, the presence or absence of the disease in question in members of the "control" group has not been ascertained. However, if the prevalence of the disease is small, then there should be hardly any difference between the effectiveness of case control design and case random design.

One important assumption underlying association studies is that the study population is **panmictic**, which means that all individuals of the opposite sex in the population are potential partners. In practice, it is often the case that random mating cannot be assumed. Within the study population there might be relatively homogeneous subgroups, for example, ethnic subgroups, social subclasses or geographically separated groups. As we will see in Sect. 5.2.3, the resulting **population structure** can have a serious effect on the statistical analysis of association studies if not accounted for appropriately.

2.2.4   Other Types of Study

This book will mainly focus on the statistical analysis of experimental populations (Chap. 4) and on panmictic populations (Chap. 5). One reason for this choice is that we will emphasize a particular approach to model selection, which has been fully developed for these two types of study in terms of both statistical theory and software. There exist a number of additional types of study. In this section, we will briefly discuss admixture mapping and some aspects of data from families. Data from families have been used extensively for linkage analysis, but more recently there has also been some interest in association studies based on such data. In this case, it seems possible to extend the statistical methods described in Chaps. 4 and 5, but the details still have to be worked out in future research.

2.2.4.1   Admixture Mapping

We mentioned above that population structures can lead to problems in association studies. However, in certain situations it can be the basis for study designs with desirable qualities. In recent years, the analysis of admixture populations has gained popularity. Such populations result from previously separated subpopulations becoming mixed. African-Americans and Latinos are perhaps the most prominent example of a pair of such subpopulations, where gene flow between these populations started only a few hundred years ago.

In admixture mapping, it is assumed that only the local LD typically found in outbred populations exists within the original subpopulations (see Fig. 2.7). From a genetical point of view, mating between two outbred populations has some similarities to the experimental crosses of inbred populations discussed in Sect. 2.2.1. Due to the process of repeated crossovers over several generations, an individual from an admixed population will have relatively long strands of DNA, where different strands stem from different ancestral populations and the length of these strands will depend on the mixing history.

The idea of admixture mapping is to use ancestral information to localize regions which influence a trait. Assume there are two founding subpopulations A and B. Then at a certain genetic locus there are three possible states of ancestry, AA, AB or BB. If a certain risk allele occurs in the ancestral population A more frequently than in B, then one would expect that the ancestral state A is observed in affected individuals from the admixed population more often at the location of that risk allele compared to other genomic regions.

One potential advantage of admixture mapping is that within recently admixed populations these strands from ancestral populations are still rather long. Hence, compared with association studies, a substantially smaller number of genetic markers are needed. Also, admixture mapping is able to locate influential regions, even when studying only cases without a control group (e.g., see [8]). On the other hand, admixture mapping only works for risk alleles which have substantially different frequencies in the ancestral populations. We recommend [18] as an introduction to such an approach and [20] to learn more about the statistical methods involved in admixture mapping.

2.2.4.2   Data from Families

As mentioned above, association studies are commonly performed based on panmictic populations and aim to find a correlation between a genetic marker and the trait in question. Data from families are normally analyzed using linkage analysis, which is based on slightly more indirect logic. In linkage studies, one tries to identify loci which cosegregate with the trait (are inherited along with the trait) within families. For a given pedigree of a family, one can compute the joint probability of specific marker genotypes and disease status. Based on such computations, one can test whether a genetic marker is in LD with a genetic locus which directly affects the trait.

An important concept in linkage analysis is being **identical by descent** (IBD). Among relatives, alleles are IBD if they arose from the same allele of a common ancestor. Tests in linkage analysis are often based on the IBD configuration, but it is not always possible to determine the IBD status of all individuals in a pedigree. This results from the problem of phasing, which is usually solved using maximum likelihood methods. However, the necessary computations become rather involved for complex pedigrees. In view of this, designs like the **affected sib-pair** method are rather popular. This is based on comparing the similarity of siblings who share the same biological mother and father. A more detailed introduction to the mathematical and statistical aspects of analyzing pedigree data can be found, for example, in [10, 15] or [16].

References

1. Broman, K.W.: The genomes of recombinant inbred lines. Genetics **169**, 1133–1146 (2005)
2. Chesler, E.J., Miller, D.R., Branstetter, L.R., Galloway, L.D., Jackson, B.L., Philip, V.M., Voy, B.H., Culiat, C.T., Threadgill, D.W., Williams, R.W., Churchill, G.A., Johnson, D.K., Manly, K.F.: The collaborative cross at Oak Ridge National Laboratory: developing a powerful resource for systems genetics. Mamm. Genome **19**, 382–389 (2008)
3. Devlin, B., Risch, N.: A comparison of linkage disequilibrium measures for fine-scale mapping. Genomics **29**, 311–322 (1995)
4. Feingold, E.: Methods for linkage analysis of quantitative trait loci in humans. Theor. Popul. Biol. **60**, 167–180 (2001)
5. Fisher, R.A.: The theory of inbreeding, 2nd edn. Academic Press, New York (1965)
6. Frommlet, F.: Tag SNP selection based on clustering according to dominant sets found using replicator dynamics. Adv. Data Anal. Classif. **4**, 65–83 (2010)
7. Haldane, J.B.S., Waddington, C.H.: Inbreeding and linkage. Genetics **16**, 357–374 (1931)
8. Hoggart, C.J., Shriver, M.D., Kittles, R.A., Clayton, D.G., McKeigue, P.M.: Design and analysis of admixture mapping studies. Am. J. Hum. Genet. **274**(5), 965–978 (2004)
9. Iraqi, F.A., Churchill, G., Mott, R.: The Collaborative Cross, developing a resource for mammalian systems genetics: a status report of the Wellcome Trust Cohort. Mamm. Genome **19**, 379–381 (2008)
10. Lange, K.: Mathematical and Statistical Methods for Genetic Analysis. Springer (1997)
11. Scheet, P., Stephens, M.: A fast and flexible statistical model for large-scale population genotype data: applications to inferring missing genotypes and haplotypic phase. Am. J. Hum. Genet. **78**, 629–644 (2006)
12. The ENCODE Project Consortium: the ENCODE (ENCyclopedia of DNA Elements) Project. Science **306**, 636–640 (2004)
13. The International Hapmap Consortium: a haplotype map of the human genome. Nature **437**, 1299–1320 (2005)
14. The International Hapmap Consortium: a second generation human haplotype map of over 3.1 million SNPs. Nature **449**, 851–862 (2007)

15. Thompson, E.A.: Pedigree Analysis in Human Genetics. Johns Hopkins University Press, Baltimore (1986)
16. Thompson, E.A.: Statistical Inferences from Genetic Data on Pedigrees. In: The Proceedings of the NSF-CBMS regional conference series in probability and statistics, vol. 6. I MS, Beachwood, OH (2000)
17. Via, M., Gignoux, C., Burchard, E.G.: The 1000 genomes project: new opportunities for research and social challenges. Genome Med. **2**, 3 (2010)
18. Winkler, C.A., Nelson, G.W., Smith, M.W.: Admixture mapping comes of age. Annu. Rev. Genomics Hum. Genet. **11**, 65–89 (2010)
19. Zhang, K., Sun, F.: Assessing the power of tag SNPs in the mapping of quantitative trait loci (QTL) with extremal and random samples. BMC Genet. **6** (2005)
20. Zhu, X., Tang, H., Risch, N.: Admixture mapping and the role of population structure for localizing disease genes. Adv. Genet. **60**, 547–569 (2008)

Chapter 3
Statistical Methods in High Dimensions

3.1 Overview

A common feature of many genetic studies today is the vast amount of data that are collected for each individual. For example, in genome wide association studies a single array provides information on the genotype of up to one million SNPs. Researchers have become accustomed to being confronted with data sets where the number of individuals n for which data have been collected is much smaller than the number of genetic markers m. However, the situation $m > n$ is very difficult to handle from a statistical point of view, and many classical approaches to multidimensional data analysis have to be reconsidered.

This chapter presents different approaches to the analysis of high-dimensional data. The first part (Sect. 3.2) is concerned with the basic concepts of multiple testing. We discuss both classical and resampling-based procedures that are designed to control the familywise error rate (FWER). Multiple testing procedures are a relatively recent development for controlling the false discovery rate (FDR). These are potentially much more powerful than procedures controlling FWER, and have been frequently adopted in applications of bioinformatics where $m > n$. Section 3.2.4 examines the notion of sparsity and presents recent results concerning the optimality of multiple testing procedures in a decision theoretic framework. It turns out that procedures controlling the FDR have somewhat similar properties to Bayesian approaches to multiple testing, which are discussed in Sect. 3.2.4.6.

The second part of this chapter gives an introduction to model selection. Information criteria based on penalized log-likelihood functions play a prominent role, and therefore we start by reviewing the concept of likelihood. Sections 3.3.2 and 3.3.3 examine the properties of two classical criteria for model selection, AIC and BIC, where in the latter section we also discuss a fully Bayesian approach to model selection. Subsequently, the main focus is on modifications of BIC which are particularly suitable for the analysis of high-dimensional data under sparsity. The chapter ends by introducing a number of other approaches to model selection, including LASSO, SLOPE, and the elastic net. These can be seen as penalized log-likelihood

© Springer-Verlag London 2016
F. Frommlet et al., *Phenotypes and Genotypes*, Computational Biology 18,
DOI 10.1007/978-1-4471-5310-8_3

methods, but in contrast to AIC and BIC their penalties depend on the magnitudes
of the estimators of the regression coefficients and not on the number of nonzero
coefficients.

This chapter is concerned with somewhat advanced statistical material, which
will be of primary importance for the applications considered in Chaps. 4 and 5. We
recommend that readers who are not entirely familiar with statistical methodology
first have a look at Chap. 8, which gives an introduction to statistical procedures
at an elementary level. Chapter 8 serves together with Chap. 6 as the foundations
on which this chapter is built up. However, readers who are sufficiently fluent with
classical statistical procedures might prefer to carry on reading and only consult the
appendices where necessary.

3.2 Multiple Testing

Multiple testing is concerned with the problem of testing more than one hypothesis
at the same time. The foundations of this subject were laid in the late 1940s and early
1950s, among others by David Duncan, Henry Scheffe, and John W. Tukey. A concise
introduction is provided by Shaffer [69]. More in-depth treatments are found in
[45, 49]. The classical theory on multiple testing presented in these books is designed
for situations where a moderate number of hypotheses are to be tested. In modern
genomic applications, the huge number of tests performed simultaneously made
it necessary to reconsider and adapt procedures for multiple testing. An excellent
review of this development is given by Dudoit et al. [30], focusing in particular on
applications in microarray data concerning gene expression. A comprehensive survey
of modern developments in multiple testing with respect to genetic applications
is [31].

To formalize the problem, consider a situation where m tests are performed simultaneously, with null hypotheses H_{0j} and alternatives H_{Aj} for $j \in \{1, \ldots, m\}$. The
number of tests for which the null hypothesis is correct is denoted by m_0. In practice,
one does not know m_0 and the statistical task is to decide which hypotheses are true
based on a sample of size n. To this end, realizations of the test statistics $T_1, \ldots, T_m$
are computed from the data sample. For the moment, we do not have to specify the
exact nature of these statistics. One might think, for example, of the t, χ^2 or F-test
statistics introduced in Chap. 8, but the following principles are rather general.

In each individual test, a choice is made between H_{0j} and H_{Aj}. Each time, there
is the possibility of a type I or a type II error. Table 3.1 summarizes the traditional
representation of this frequentist setting using the notation of [4], where V is the
number of false positives and T the number of false negatives.

An important question is how to generalize the concept of type I error rates
to multiple hypothesis testing. Definitions of the most traditional measures can be
found in the introduction of [45]. The per-comparison error rate, PCER $= E(V)/m$,
is defined as the expected proportion of type I errors, whereas the per-family error
rate, PFER $= E(V)$, is the expected number of type I errors. Historically, the most

Table 3.1 Notation for multiple testing

	Number of non-rejections	Number of rejections	Total
True null hypotheses	U	V	m_0
True alternatives	T	S	$m - m_0$
	$m - R$	R	m

important concept in multiple testing used to be the familywise error rate, FWER = $P(V > 0)$, which is the probability of making any type I error within a family of m tests. From the chain of inequalities

$$\frac{1}{m} \sum_{j=1}^{m_0} P(B_j) \leq \max_{j \in \{1,\dots,m_0\}} P(B_j) \leq P \left(\bigcup_{j=1}^{m_0} B_j \right) \leq \sum_{j=1}^{m_0} P(B_j), \qquad (3.1)$$

it immediately follows that PCER $\leq$ FWER $\leq$ PFER. The last inequality above is Boole's inequality, which is sometimes also called the first-order Bonferroni inequality. Note that the concept of PCER does not address the multiple testing problem at all. As long as all individual tests control the type I error rate at level α, then PCER $\leq \alpha$.

Controlling a generalized type I error rate is one objective. Ideally, one would also like to minimize a generalization of the type II error. For example, one might like to maximize the average power $E(S)/m$, which is the average probability of detecting a true signal. In general, it is difficult to find a good compromise between controlling the number of false positives and maximizing the power. In particular, when m is very large, controlling FWER leads to procedures that tend to miss a lot of true signals and thus have a rather low average power. For this reason, recently many less stringent type I error control rates have been introduced, among them the generalized familywise error rate gFWER(k) = $P(V > k)$, which is the probability of making at least k type I errors. One of the most popular recent innovations, introduced by Benjamini and Hochberg [4], is the false discovery rate (FDR) defined below in Sect. 3.2.3. A comprehensive overview of other generalized type I error rates is given by [31]. For example, [54] considers procedures controlling the number of false discoveries rather than the FDR.

3.2.1 Classical Procedures Controlling FWER

In general, the test statistics $T_1, \dots, T_m$ form a random vector that might have a rather complicated dependence structure. The simplest multiple testing procedures (MTP) do not take such dependence into account and are only based on the marginal distributions. These procedures can be understood in terms of the p-values p_j of individual tests. The best known and most widely used MTP in the applied sciences is commonly referred to as the Bonferroni correction.

The Bonferroni procedure:

Let $\alpha_j > 0$, $j = 1, \ldots, m$ be positive real numbers with $\sum_{j=1}^{m} \alpha_j = \alpha$.

Then we reject H_{0j} whenever $p_j \leq \alpha_j$.

Often, the standard choice of $\alpha_j = \alpha/m$ is made, where α is the so-called nominal significance level for the MTP. When the T_j are continuous random variables, the p values under H_{0j} are uniformly distributed, $p_j \sim \mathscr{U}(0, 1)$ (see Sect. 6.3.3 for the definition of the uniform distribution). The expected number of type I errors is $\sum_{j=1}^{m_0} P(p_j \leq \alpha_j)$, where

$$\sum_{j=1}^{m_0} P(p_j \leq \alpha_j) \leq \sum_{j=1}^{m_0} \alpha_j \leq \alpha.$$

Therefore, it immediately follows that the Bonferroni procedure controls PFER at the level α. According to Boole's inequality (3.1), FWER is also controlled at the nominal level α. The following MTP is closely related to the Bonferroni rule.

The Šidák procedure:
Reject H_{0j} whenever $p_j \leq 1 - (1 - \alpha)^{1/m}$.

This correction for multiple testing is based on Šidák's inequality

$$P\left(\bigcap_{j=1}^{m_0}(p_j > \alpha_j)\right) \geq \prod_{j=1}^{m_0} P(p_j > \alpha_j), \tag{3.2}$$

which holds under fairly general conditions (see [31] for the details). One might also define Šidák's correction with unequal allocations α_j, such that $\prod (1 - \alpha_j) = (1 - \alpha)$. However, the simplest and most common choice is $\alpha_j = 1 - (1 - \alpha)^{1/m}$. In the case of independent test statistics, (3.2) holds with equality, and under the total null (i.e., when $m = m_0$), Šidák's procedure controls FWER exactly at the nominal level α. Since $1 - (1 - \alpha)^{1/m} \geq \alpha/m$, Šidák's procedure is slightly more powerful than the Bonferroni correction. However, Fig. 3.1 shows that in the case of independent test statistics the difference is extremely small.

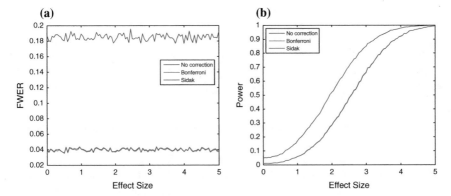

Fig. 3.1 Power and FWER for the Bonferroni and Šidák multiple testing procedures, and without correction for multiple testing. Data were generated for five independent z-test statistics, one of which was simulated under the alternative hypothesis with varying effect sizes as shown on the x-axis of both plots. Clearly, correction for multiple testing is necessary to control the family-wise error rate at a prespecified level α. Note that both the Bonferroni and the Šidák procedures were performed at the nominal level $\alpha = 0.05$. The actual significance level is really 0.04, due to the fact that only four out of the five null hypotheses are correct. For this example, both procedures are conservative

Given a particular joint distribution of the test statistics $T_1, \ldots, T_m$, the true FWER of an MTP is called its actual significance level. An MTP is referred to as being conservative when its actual significance level is smaller than the nominal level α. Both the Šidák and Bonferroni procedures are often quite conservative, in particular, when the test statistics are positively correlated. A great number of more refined MTPs that control FWER have been developed to deal with such situations. A thorough treatment of classical scenarios, such as comparison of treatment means, is given by [45, 49]. More recent developments, like the use of resampling methods to take into account unknown correlation structures, are discussed in Sect. 3.2.2.

The Bonferroni procedure controls FWER for any given number of true null hypotheses m_0, a property that is referred to as controlling the type I error rate in the strong sense. A procedure that controls FWER when all the null hypotheses are true, i.e. $m = m_0$, is said to control the type I error rate in the weak sense. However, in the case when m_0 is substantially smaller than m, the Bonferroni procedure tends to be very conservative—the actual FWER is substantially smaller than α. This results in an unnecessary loss of power. The Holm–Bonferroni procedure [48] is another method that controls FWER in the strong sense, but achieves greater power than the Bonferroni correction. This results from applying the closed testing principle [59], which is a powerful tool for constructing procedures that control FWER.

To describe the Holm–Bonferroni procedure, let $p_{[1]} \le p_{[2]} \le \cdots \le p_{[m]}$ be the ordered p values of the m individual tests and $H_{0[j]}$ the corresponding ordered null hypotheses.

The Holm–Bonferroni step-down procedure:

Let k be the smallest $j \in \{1, \ldots, m\}$ for which $p_{[j]} > \frac{\alpha}{m+1-j}$.

If no such k exists, then reject all the null hypotheses, otherwise reject all $H_{0[j]}$ with $j < k$.

The Bonferroni and Šidák corrections are so-called single step MTPs, i.e., there is one corrected significance level, according to which each inference is made. In contrast, the Holm–Bonferroni procedure is a stepwise MTP, i.e., rejection or acceptance of a particular null hypothesis T_{0j} may depend on the values of the other test statistics T_{0i}, $i \neq j$. The benefit gained from adopting the stepwise Holm–Bonferroni MTP is that, compared with the simple Bonferroni correction, it has greater power (see Fig. 3.2). In complete analogy to the Holm–Bonferroni MTP, one can devise a stepwise MTP based on the Šidák inequality. One drawback of stepwise MTPs is the fact that the rejection regions do not correspond to confidence intervals related to individual tests, which is the case for single step MTPs.

The Holm–Bonferroni procedure belongs to a class of MTPs which are usually referred to as step-down procedures. One starts with the hypothesis having the smallest p value and sees whether it can be rejected at the Bonferroni significance level α/m. If not, one accepts all the null hypotheses, otherwise one goes one step further and sees whether the hypothesis with the second smallest p value can be rejected at the significance level $\alpha/(m-1)$. One continues until it is not possible to reject $H_{0[k]}$ at the significance level $\alpha/(m-k+1)$ for the first time. Each step-down procedure has a corresponding step-up procedure. Using a step-up procedure, one starts

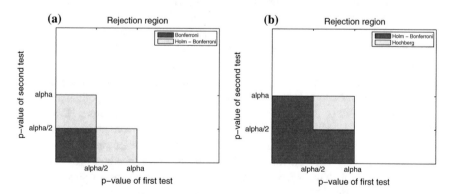

Fig. 3.2 Rejection regions for multiple testing procedures when two hypotheses are tested. The *red* area corresponds to the sets of p values where both procedures reject both null hypotheses. The *yellow* area corresponds to the sets of p values where only the procedure with greater power rejects both null hypotheses. Plot **a**: The Holm–Bonferroni procedure has a larger rejection region, and therefore also greater power, than the single step Bonferroni procedure. Plot **b**: On the other hand, the Hochberg procedure has an even larger rejection region than the Holm–Bonferroni procedure

by checking the hypothesis corresponding to the largest p value. If one can reject it at the appropriate significance level, one rejects all the null hypotheses, otherwise one accepts $H_{0[m]}$ and continues on to the next largest p value until the first rejection occurs. The step-up procedure corresponding to the Holm–Bonferroni MTP was introduced by Hochberg [46].

The Hochberg step up procedure:

Let k be the largest $j \in \{1, \ldots, m\}$ for which $p_{[j]} \leq \frac{\alpha}{m+1-j}$.

Then we reject all $H_{0[j]}$ with $j \leq k$.

A step-up MTP is always more powerful than its corresponding step-down MTP, due to the fact that it has a larger rejection region. However, it is more difficult for step-up procedures to control a type I error rate. The justification for using the Hochberg step-up MTP to control FWER is based on the Simes inequality

$$P\left(p_{[j]} > \frac{j\alpha}{m}, \ \forall j = 1, \ldots, m | m_0 = m\right) \geq 1 - \alpha, \tag{3.3}$$

which holds for independent test statistics, as well as for a variety of dependency structures (for details see e.g. [31]). Hochberg showed in [46] that his step-up procedure controls FWER whenever (3.3) holds.

Recently, there has been some interest in procedures that control generalizations of the FWER based on the probability of making at least k false positives, gFWER(k). Lehmann and Romero [56] studied Bonferroni type corrections controlling gFWER(k) using the corrected significance level $\frac{k\alpha}{m}$, as well as the corresponding step-down MTP.

3.2.2 Permutation Tests and Resampling Procedures

The multiple testing procedures presented in Sect. 3.2.1 are all based on p values and therefore rely upon the appropriate assumptions regarding the distributions of the test statistics. For example, the two sample t-test (8.4) assumes that the data come from two normally distributed populations. In this case, the appropriate t-test based on the Student distribution is the most efficient (see [63]). If these assumptions are seriously violated, then classical nonparametric tests can be used, as described in Sect. 8.7. Donoho and Tanner [29] carry out an analysis of the effect of deviations from normality on inference for large data sets.

The nonparametric counterpart of the two sample t-test is the Wilcoxon rank-sum test (8.23), which is based on the distribution of the ranks of the observations in the two groups. P values for the rank-sum test are obtained using the following combinatorial considerations: Under the null hypothesis, which states that the data from both groups have the same underlying probability distribution, the distribution of the ranks in the two groups is a discrete uniform distribution. This means that each possible constellation of ranks has exactly the same probability, namely one divided by the number of possible permutations of the observations (see [58] for details).

If the numbers of group members are n_1 and n_2, respectively, then under the null hypothesis the probability of a specific constellation of ranks equals $1/n!$, where $n = n_1 + n_2$ is the total number of observations. The p value is then simply obtained by counting the number of all constellations that are 'more extreme' than the observed one, which means that they have a larger realization of the test statistic. For small n, this procedure is simple and can be easily done by hand. For larger n, the number of permutations grows so fast that it is too demanding to consider all possible permutations, even for modern computers. However, if n is large then it is not really necessary to consider all permutations to obtain good approximations of p values, but it suffices to consider a sufficiently large random subsample of permutations. The proportion of sampled permutations with a larger realization of the test statistic than the observed one serves as an estimate of the p value. Typically, around $B = 10000$ or more permutations are performed to obtain a reasonably precise estimate.

Resampling-based testing clearly has the advantage of providing p values without making any distributional assumptions on the test statistic. In the context of multiple testing, there is a second benefit: Whereas the methods discussed in Sect. 3.2.1 fail to incorporate the dependence structure between different test statistics, resampling schemes can address this issue, at least to some extent. Here, we discuss the basic ideas in the context of a relatively simple example. We consider only permutation tests, which means that we sample from our data without replacement. The classical book on resampling-based multiple testing is [77], where one can also find extensive treatment of the bootstrap approach, which means resampling with replacement. We also particularly recommend [31], which provides a very concise synopsis of permutation tests in the context of microarray analysis (see also [25] and [54]).

Let us consider m statistical tests performed to decide whether m different variables show any difference in their location parameters between two groups. These tests are based on the data from n individuals, x_{ij}, with $i = 1, \ldots, n$ and $j = 1, \ldots, m$. Suppose the variable Y codes the group membership. Assume that the first n_1 individuals belong to the group $Y = 1$ and the following n_2 individuals belong to the group $Y = 2$. Furthermore, the data from individuals in the first group $x_{i1}, \ldots, x_{im}$, $i \leq n_1$ are thought of as independent realizations from a random vector $\mathbf{X}^{(1)} = (X_1^{(1)}, \ldots, X_m^{(1)})$ and, similarly, the data from the second group are given by a random vector $\mathbf{X}^{(2)} = (X_1^{(2)}, \ldots, X_m^{(2)})$. The data can be summarized as follows:

Y	$X_1^{(Y)}$	$X_2^{(Y)}$	$\cdots$	$X_m^{(Y)}$
1	x_{11}	x_{12}	$\cdots$	x_{1m}
$\vdots$	$\vdots$	$\vdots$		$\vdots$
1	$x_{n_1 1}$	$x_{n_1 2}$	$\cdots$	$x_{n_1 m}$
2	$x_{(n_1+1)1}$	$x_{(n_1+1)2}$	$\cdots$	$x_{(n_1+1)m}$
$\vdots$	$\vdots$	$\vdots$		$\vdots$
2	x_{n1}	x_{n2}	$\cdots$	x_{nm}

Let $\mathbf{X}^{(1)}$ have a joint probability distribution with marginal distribution functions $F_j^{(1)}$, $j = 1, \ldots, m$ and, similarly, let the second group have marginal distribution functions $F_j^{(2)}$, $j = 1, \ldots, m$. Assume that the marginal distributions only differ according a location parameter, that is, $F_j^{(2)}(x) = F_j^{(1)}(x - a_j)$. We thus want to test the following m hypotheses:

$$H_{0j} : a_j = 0, \quad \text{against} \quad H_{1j} : a_j \neq 0, \quad j = 1, \ldots, m. \tag{3.4}$$

This is a multivariate generalization of the setting where a Wilcoxon rank sum test is appropriate. However, using permutation tests, one can be quite flexible with respect to the test statistics used. Hence, one can also work directly with the common two sample t-test statistic. So the first step is to compute the test statistics based on the observed data:

$$t_j = \frac{\bar{x}_j^{(1)} - \bar{x}_j^{(2)}}{s_p}, \quad j = 1, \ldots, m, \tag{3.5}$$

with $\bar{x}_j^{(1)} = \frac{1}{n_1} \sum_{i=1}^{n_1} x_{ij}$, $\bar{x}_j^{(2)} = \frac{1}{n_2} \sum_{i=n_1+1}^{n} x_{ij}$, and

$$s_p^2 = \left(\frac{1}{n_1} + \frac{1}{n_2} \right) \frac{\sum_{i=1}^{n_1} (x_{ij} - \bar{x}_j^{(1)})^2 + \sum_{i=n_1+1}^{n} (x_{ij} - \bar{x}_j^{(2)})^2}{n_1 + n_2 - 2}.$$

We denote the corresponding random variables underlying the observed test statistics as T_j, $j = 1, \ldots, m$. Next comes the resampling step, which starts by performing a random permutation $\pi(Y)$ of the vector of group membership. The matrix x_{ij} remains unchanged, which means that the dependence structure between the different variables remains unchanged. Then the t-test statistics are computed according to Eq. (3.5) for the shuffled groups given by $\pi(Y)$. This resampling step is repeated B times and each permutation yields a vector of test statistics $\mathbf{t}^{(b)} = (t_1^{(b)}, \ldots, t_m^{(b)})$, with $b = 1, \ldots, B$.

If the alternative hypothesis H_{1j} holds for a certain variable, then there is a large difference between $E(X_j^{(1)})$ and $E(X_j^{(2)})$. Accordingly, one would also expect to observe a large difference between $\bar{x}_j^{(1)}$ and $\bar{x}_j^{(2)}$ in the original data, which results in the realization of the test statistic t_j having a large absolute value. Now for the reshuffled group memberships given by $\pi(Y)$, individuals are randomly assigned to a group. Therefore, there is no association between $\pi(Y)$ and x_{ij}, thus one cannot expect to see any difference between the group averages. Thus the realizations of the resampling based test statistics $\mathbf{t}^{(b)}$, $b = 1, \ldots, B$ can be seen as a sample obtained under the global null hypothesis $a_1 = \cdots = a_m = 0$. This sample can then be used to define critical values for $(T_1, \ldots, T_m)$, which allow us to control the FWER.

Consider

$$M^{(b)} := \max(|t_1^{(b)}|, \ldots, |t_m^{(b)}|),$$

i.e., $M^{(b)}$ is the maximum of the m realizations of the absolute value of the test statistic for the b-th resample. For sufficiently large B, the empirical distribution function of the sample $M^{(b)}$, $b = 1, \ldots, B$ will be very close to the distribution of the maximum of the absolute value of m test statistics $(T_1^0, \ldots, T_m^0)$, which stem from the global null hypothesis (i.e., no shift in location) and have a similar correlation structure to the random variables $(T_1, \ldots, T_m)$ underlying the test statistics for the original data. Now the key idea is to define critical values based on the quantiles of the resampled maxima. One defines C to be the $1 - \alpha$ quantile of $M^{(b)}$, $b = 1, \ldots, B$. The j-th null hypothesis H_{0j} is thus rejected if $|t_j| > C$. It can be shown that this procedure controls FWER in the strong sense.

This procedure is rather intuitive. It simply compares the observed realizations of the test statistics with the maxima of the realizations of the test statistics generated under the global null hypothesis. This is the same reasoning that leads, in the case of independent test statistics, to Šidák's procedure. However, if the test statistics are dependent, then it is usually not possible to obtain an analytic expression for the distribution of the maximum of the m statistics. This resampling procedure can be thought of as a computer intensive method of approximating the distribution of this maximum under the global null hypothesis. Furthermore, it allows us to obtain approximations of the so-called adjusted p values, which are defined as

$$\tilde{p}_j := P\left(\max_{1 \leq j \leq m} |T_j^0| \geq |t_j|\right), \tag{3.6}$$

where the T_j^0 are the random variables describing the test statistics under the global null hypothesis and the t_j are the observed realizations of the test statistics, $j = 1, 2, \ldots, m$. Based on resampling, the adjusted p value $\tilde{p}_j$ can simply be estimated as the proportion of maxima $M^{(b)}$ whose absolute value is larger than $|t_j|$.

We have presented here only the simplest single step adjustment, which can be seen as the resampling counterpart of the Bonferroni (or rather the Šidák) procedure. As mentioned above, an excellent source for a more in-depth treatment of resampling-based multiple testing is [77], where one will also find a detailed description of stepwise resampling procedures (for example, the analogue of the Holm–Bonferroni procedure). Permutation tests are a widely used tool in genetic association studies, and in view of the general topic of this book we would like to mention [25, 28], which introduced permutation tests in the context of QTL mapping.

3.2.3 Controlling the False Discovery Rate

One of the biggest recent innovations in multiple testing was the proposal by Benjamini and Hochberg [4] to make use of the false discovery rate (FDR). The FDR is formally defined as

$$\text{FDR} = E(V/R), \quad \text{with } V/R = 0 \text{ in the case of } R = 0, \tag{3.7}$$

or equivalently $\text{FDR} = P(R > 0)E(V/R|R > 0)$. This type I error rate is designed to control the expected proportion of incorrectly rejected null hypotheses. Under the total null hypothesis, i.e., $m_0 = m$, the FDR is equivalent to the FWER. Therefore, an MTP that controls the FDR also controls the FWER in the weak sense. Otherwise, FDR control involves less stringent control of the type I error rate than FWER control, resulting in a potential gain in power. In [4], Benjamini and Hochberg discussed the False Discovery Rate of the following stepwise procedure:

Benjamini-Hochberg step up procedure (BH):

Let k be the largest $j \in \{1, \ldots, m\}$ for which $p_{[j]} \leq \frac{j\alpha}{m}$.

Then we reject all $H_{0[j]}$ with $j \leq k$.

In [4], it is shown that if the test statistics corresponding to different tests are independent, then the FDR of this procedure is equal to $\frac{m_0}{m}\alpha$, where m_0 is the number of true null hypotheses.

The above stepwise procedure had been proposed in a series of papers by Eklund in 1961–1963 (see [68]) and later by Simes [70] as a test for the total null hypothesis. However, due to the fact that BH does not control FWER in the strong sense, its use was never quite accepted before the work of Benjamini and Hochberg. For this reason, the procedure is today generally referred to as the Benjamini–Hochberg procedure, a convention which we adopt. Note that in comparison to the Holm–Bonferroni procedure, the threshold levels for p values are considerably larger according to BH. The resulting gain in power, together with the fact that BH is really easy to apply, has made this MTP fairly popular in the applied sciences, in particular in the field of bioinformatics. On the other hand, [4] has inspired a large amount of theoretical work. A deeper understanding of BH from a statistical point of view is provided in [36].

The Benjamini–Hochberg procedure is a step-up procedure and, consequently, there exists an analogous step-down procedure (SD) that controls FDR. Figure 3.3 illustrates the difference between BH and SD, where BH selects the alternative three times, whereas SD does so only once. While in general BH is more powerful than SD, in practice there are hardly any differences.

The original work of Benjamini and Hochberg [4] only dealt with the situation of independent tests. Later, Benjamini and Yekutieli [6] proved that BH also controls the

Fig. 3.3 Example illustrating the difference between the Benjamini–Hochberg procedure (BH) and its corresponding step-down procedure (SD) at an FDR level of $\alpha = 0.5$. Based on the p values for 200 hypotheses (of which only 20 are shown in this graph), SD finds only one alternative significant, whereas the more powerful BH procedure selects three

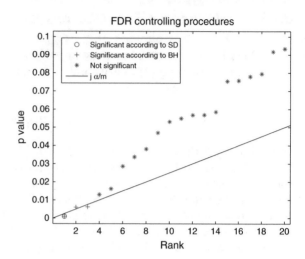

false discovery rate when the test statistics T_j corresponding to the null hypothesis have a positive regression dependency. They also proposed a conservative modification of BH, which controls FDR for any type of dependency between test statistics. The work of Storey [72] on the positive false discovery rate, $E(V/R|R > 0)$ is closely related, as are the procedures introduced by Lehmann and Romano [56] to control the quantiles of the false discovery proportion V/R.

3.2.4 Multiple Testing Under Sparsity. Minimizing the Bayesian Risk in Multiple Testing Procedures

This section presents some more advanced methods for multiple testing and a discussion of their theoretical properties under sparsity. In comparison to the methods discussed so far, this material is substantially more advanced. We still encourage the interested reader to study this section, since in many practical situations, more sophisticated statistical methods enable more efficient analysis of complex high-dimensional data sets. Also, an understanding of some basic theoretical properties of these methods enables us to choose the statistical methodology that is best suited to the character of a given set of data and the purpose of the statistical analysis.

Sparsity is a key notion in multiple testing when the number of tests is larger than the sample size. It signifies that the number of true alternative hypotheses, $m_A = m - m_0$, is much smaller than the total number of tests. In this section, the proportion of alternative hypotheses that are true is denoted by $p = m_A/m$. Small values of p indicate sparsity.

3.2.4.1 Estimation of Subpopulation Means

Apart from the previously discussed standard measures of type I and type II error rates, one can also adopt a decision theoretic approach and evaluate multiple testing procedures with respect to the expected value of some loss function. The seminal paper [1] considered the problem of estimating the means μ_j in a normal population $T_j \sim N(\mu_j, \sigma^2)$, $j = 1, \ldots, m$, under the additional assumption that the means differ from zero only in a small proportion, denoted p, of cases. Let us denote the vector of population means by $\mu = (\mu_1, \ldots, \mu_m)$ and the vector of their estimates by $\hat{\mu} = (\hat{\mu}_1, \ldots, \hat{\mu}_m)$.

The main focus of [1] was on the accuracy of estimating μ, which can be measured, for example, by $L^r(\hat{\mu}, \mu) = \sum_{j=1}^{m} |\hat{\mu}_j - \mu_j|^r$ for some $r > 0$. The natural estimate of μ_j seems to be T_j. However, it is well known that for $m > 2$, the vector of test statistics $T_1, \ldots, T_m$ is suboptimal with respect to minimizing the mean squared error $E\left(L^2(\hat{\mu}, \mu)\right)$ and can be improved, for example, by the James–Stein estimator

$$\hat{\mu}_j^{JS} = \left(1 - \frac{(m-2)\sigma^2}{||T||^2}\right) T_j,$$

where $||T||^2 = \sum_{j=1}^{m} T_j^2$ (see [50, 71]). The James–Stein estimator shrinks all test statistics towards 0.

A different approach is suggested in [1], where sufficiently small T_j are simply replaced by 0. Specifically, [1] uses the BH procedure to decide when to set $\hat{\mu}_j = 0$, and it is shown that when $p \in \left[\frac{\log^5 m}{m}, m^{-\delta}\right]$ for some $\delta > 0$, then this hard thresholding procedure has some asymptotic optimality properties with respect to minimizing $E\left(L^r(\hat{\mu}, \mu)\right)$ when $r \in (0, 2]$. A necessary condition is that the sequence of nominal FDR levels for BH, α_m, satisfy $\alpha_m \geq \frac{\gamma}{\log m}$ for some $\gamma > 0$ and $\lim_{m \to \infty} \alpha_m = C \in [0, 1/2]$. Note that the above condition on p implies that the sparsity levels converge towards zero ($p \to 0$) at such a rate that the number of true signals (alternatives) increases with m.

The results of [1] show that for a wide range of sparsity levels p, the BH procedure adapts well to the unknown degree of sparsity, and for large m, BH gives near optimal precision when estimating the vector of means. Moreover, [1] also considers other definitions of sparsity, where all the μ_j are different from zero, but the majority of them are very small. BH also possesses similar optimality properties in such a situation.

3.2.4.2 Classification Problem

In [9, 12, 39, 42, 62], the behavior of classical MTPs in the context of sparsity is discussed with respect to a standard loss function used for classification problems. Specifically, [9, 12, 39] consider the standard problem of testing hypotheses about means μ_j in normal populations $T_j \sim N(\mu_j, \sigma^2)$, $j = 1, \ldots, m$. The starting point is a model with two groups [35], which is discussed in more detail in Sect. 3.2.4.6. The simplest model covered by the analysis of [9, 12] is the following mixture model:

$$T_j \sim (1 - p)\mathcal{N}(0, \sigma^2) + p\mathcal{N}(0, \sigma^2 + \tau^2), \qquad (3.8)$$

which corresponds to testing the hypotheses

$$H_{0j} : \mu_j = 0 \quad \text{against} \quad H_{Aj} : \mu_j \sim \mathcal{N}(0, \tau^2). \qquad (3.9)$$

This is similar to the classical two-sided test, but in each test under the alternative the effect size is a random variable. The model given by (3.8) assumes that the true signals μ_j have a normal $N(0, \tau^2)$ distribution.

In this presentation, we assume that σ is known. This assumption can be justified by the fact that in many practical applications T_j is the average of a sufficient number of observations to reasonably estimate σ. Some solutions that can be applied in the case of a lack of replicates are discussed in the following sections.

Consider m independent tests based on the model (3.8). A loss function is defined such that the loss is δ_0 for a type I error and δ_A for a type II error. The total loss is defined to be the sum of losses in the individual tests. For a particular MTP, let t_{1j} and t_{2j} be the probabilities of type I and type II errors for the j-th hypothesis, respectively. Then the corresponding Bayes risk is defined as the expected total risk

$$R = m \left((1 - p)t_{1j}\delta_0 + pt_{2j}\delta_A\right). \qquad (3.10)$$

For the special case $\delta_0 = \delta_A = 1$, this is simply the expected number of misclassifications.

Fig. 3.4 The dependency of the expected number of misclassified hypotheses on the sparsity parameter p, $m = 200$, $\sigma = 1$, $\tau = \sqrt{2 \log 200} \approx 3.26$

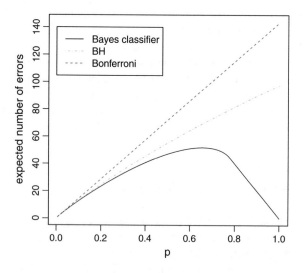

For a mixture model (3.8), the Bayes rule minimizing (3.10) is given by the Bayes classifier (see Sect. 9.4), which rejects the hypothesis H_{0j} if and only if

$$\frac{T_j^2}{\sigma^2} > \frac{\sigma^2 + \tau^2}{\tau^2} \left(\log \left(\left(\frac{\tau}{\sigma}\right)^2 + 1 \right) + 2 \log \left(\frac{1-p}{p} \frac{\delta_0}{\delta_A} \right) \right). \tag{3.11}$$

Figure 3.4 illustrates the dependency of the expected number of misclassified hypotheses on the sparsity parameter p for the Bayes classifier based on $\delta_0 = \delta_A = 1$, BH and the Bonferroni rule. As discussed in [12, 42], for very small p the risk when using the Bonferroni procedure or BH is similar to the risk when using the Bayes classifier. It is also important to note that the range of small values of p for which these procedures work well with respect to minimizing the Bayes risk is substantially larger for BH than for the Bonferroni procedure. These empirical results are confirmed by the theoretical results of [9], briefly described below.

In asymptotic theory, sparsity is often formally modeled by letting the fraction of true alternative hypotheses p decrease to 0 as the number of tests m increases to infinity. In particular, a common choice is to assume $p \propto m^{-\beta}$ for some $\beta \in (0, 1]$, where β describes the asymptotic level of sparsity. The asymptotic theory derived in [9] includes two further parameters. First, the strength of the signal under the alternative, $u = \tau/\sigma$, is assumed to be proportional to $\log m$. In [9], such signals are referred to as being on "the verge of detectability," since the asymptotic power of the corresponding optimal Bayes classifier is larger than 0 and smaller than 1. Such signals are very interesting, since they are at the level of noise and enable us to choose appropriate test procedures. The second of the additional parameters is the ratio between the losses for type I and type II errors, $\delta = \delta_0/\delta_A$.

Let us denote the risk under the Bayes rule by R_{opt}. Under the asymptotic assumptions described above, an MTP is classified as asymptotically Bayes optimal under sparsity (ABOS) if the ratio between the corresponding Bayes risk and the risk under the optimal Bayes classifier converges to one, i.e., $R/R_{opt} \to 1$. According to [9], the following optimality results hold:

Optimality results for Bonferroni and Benjamini–Hochberg procedures:

Consider the asymptotic framework

1. $m \to \infty$, $p \propto m^{-\beta}$ for $\beta \in (0, 1]$
2. $\log \delta = o(\log m)$
3. $u \propto \log m$.

Under the additional conditions

$$\log \alpha = o(\log m) \quad \text{and} \quad \alpha\delta \to 0,$$

the Bonferroni correction is ABOS only for $\beta = 1$, while BH is ABOS for any $\beta \in (0, 1]$.

Thus, in contrast to [1], in the context of classification loss, the asymptotic optimality of BH also holds for $p \propto 1/m$. This is the assumption of "extreme" sparsity, under which the expected number of true signals does not increase with m. This makes it difficult to accurately estimate both p and the distribution of effect sizes under the alternative. Interestingly, the results of [9] also show that the Bonferroni correction has some asymptotic optimality properties under such "extreme" sparsity.

The results of [9] were further extended by [62] to cover a variety of tests for location and scale parameters. Moreover, in [39] it is shown that if the sample size used to calculate each of the test statistics slowly increases with m, then the results of [9] can be generalized to alternatives where the distribution of the true effect sizes has any positive bounded density over the real line.

3.2.4.3 Full Bayes Approach

In this and the following section, based on the full Bayes and empirical Bayes approaches, we present analyses of the model (3.8) specifically designed for the situation of sparsity. A brief general introduction to Bayesian analysis is given in Chap. 9.

To apply a full Bayes approach, we need to define some prior distribution for the unknown parameters τ and p. For the purpose of illustration, we use a noninformative prior for τ^2, suggested in [66], which has density function

$$\pi_1(\tau^2 | \sigma^2) = \frac{\sigma^2}{(\sigma^2 + \tau^2)^2}. \tag{3.12}$$

With regard to the parameter p, it was observed in [12] that under sparsity a noninformative uniform prior leads to a large number of false rejections. Therefore, it was suggested to use the following informative Beta prior on p, which is concentrated on small values of p, i.e., assumes some form of sparsity:

$$\pi_2(p) = \beta(1 - p)^{\beta - 1}, \tag{3.13}$$

with $\beta > 1$. For the specific choice $\beta = 22.76$, the median of (3.13) is equal to 0.03. This corresponds, for example, to a situation where one has $m = 200$ hypotheses to be tested and in advance expects to detect roughly six signals different from zero.

In the case when σ is unknown, it is common to consider the following noninformative improper prior on this parameter: $\pi_0(\sigma^2) = \frac{1}{\sigma^2}$ (see [12, 66]).

To compute the posterior probability of H_{0j}, we introduce the following dummy variables: $\gamma_j = 0$ if H_{0j} holds and $\gamma_j = 1$, otherwise. Then the posterior probability that the j-th hypothesis is not rejected, $P(H_{0j} | T_1, \ldots, T_m) = P(\gamma_j = 0 | T_1, \ldots, T_m)$, can be estimated via the Metropolis–Hastings algorithm(see Sect. 9.3.2). Classification based on the posterior probabilities obtained from a full Bayes analysis is usually performed by rejecting the hypothesis H_{0j} if and only if

$$P(H_{0j}|T_1, \ldots, T_m) < 0.5. \tag{3.14}$$

3.2.4.4 Parametric Empirical Bayes

The parametric empirical Bayes approach (see Sect. 9.3.4) is very popular in the context of analyzing microarray data. Here, one typically has a rather small number of replicates and, consequently, the standard maximum likelihood estimates of the variance terms for the individual gene expression levels are rather imprecise. Hierarchical Bayesian models are often used as an alternative approach and use the same prior distribution for the standard deviation of each of the test statistics. Hence, the parameters of this distribution and the standard deviation of each of the respective t-statistics can be estimated based on the empirical Bayes approach. This allows us to borrow information from different test statistics to estimate the individual variance terms and usually increases the precision of these estimates. The method is implemented in the LIMMA package [78], which has become the standard tool for practical microarray analysis. In this section, we discuss a different parametric empirical Bayes method that allows us to construct an empirical version of the Bayes classifier in the setup specified by model (3.8).

The natural way of applying the Bayes classifier (3.11) in the situation when the parameters of the mixture model (3.8) are unknown is to use consistent estimators and plug them into (3.11). Based on this idea, in [12] the following parametric empirical Bayes procedure was proposed. Let

$$L(T_1, \ldots, T_m|p, \tau) = \prod_{i=1}^{m} (p f_A(T_i) + (1 - p) f_0(T_i)),$$

where $f_0(x)$ and $f_A(x)$ are the density functions of the normal distributions $\mathcal{N}(0, \sigma^2)$ and $\mathcal{N}(0, \sigma^2 + \tau^2)$, respectively.

The parameters of the model, p and τ, are estimated in two steps. First, p is fixed and $\tau(p)$ is estimated using the method of moments. In [12], the fourth moment of the mixture distribution was applied, which makes the procedure very sensitive to changes in the tail and hence yields good results in the case of very sparse mixtures.

Let us denote the fourth moment of T_j by μ_4. It is easy to check that

$$\mu_4 = 3\sigma^4 + 3p\tau^4 + 6p\sigma^2\tau^2.$$

Thus,

$$\tau^2 = \sqrt{\frac{\mu_4 - 3\sigma^4}{3p} + \sigma^4} - \sigma^2.$$

Plugging in the fourth moment of the sample $\alpha_4 = \frac{\sum_{j=1}^m T_j^4}{m}$ in place of μ_4, we obtain the following estimate of $\tau(p)$:

$$\hat{\tau}(p) = \begin{cases} \sqrt{\frac{\alpha_4 - 3\sigma^4}{3p} + \sigma^4} - \sigma^2 & \text{if } \alpha_4 > 3\sigma^4 \\ 0 & \text{if } \alpha_4 \leq 3\sigma^4. \end{cases}$$

If $\alpha_4 \leq 3\sigma^4$, then there is a strong indication that the data do not contain observations from the alternative distribution, in which case $\hat{p} = 0$. Otherwise, p is estimated by maximizing

$$\hat{p} = \text{argmax}_p \{\log L(T_1, \ldots, T_m | p, \hat{\tau}(p)) - (\beta - 1) \log(1 - p)\} \qquad (3.15)$$

using numerical methods. Here, β is the parameter of the prior distribution for p specified in (3.13) and the resulting estimator of p can be interpreted as the mode of the "posterior" density. Finally, the estimators $\hat{p}$ and $\hat{\tau}(\hat{p})$ are plugged into (3.11). The resulting parametric empirical Bayes classifier is denoted by PEB.

Again, this method can be easily adapted to the case where σ is unknown by jointly estimating $\sigma(p)$ and $\tau(p)$ using the second and fourth moments of the sample of T.

3.2.4.5 Comparison of Different Methods

We now want to compare the performance of the Bayes and empirical Bayes approach with the Benjamini–Hochberg procedure BH and with a modification thereof. It was pointed out in Sect. 3.2.3 that, for independent test statistics, the FDR of BH is equal to $\frac{m_0}{m}\alpha$. When m_0 is known, one can thus easily modify BH to control the FDR at the level α by replacing k with $k1 = \text{argmax}_j \left\{ p_{[j]} \leq \frac{j\alpha}{m_0} \right\}$.

In [5], a graphical method of estimating m_0 is proposed and a corresponding adaptive version of BH is constructed. Furthermore, in [34] it was observed that under the two groups mixture model, like the one defined in (3.8), BH with k replaced by

$$k2 = \text{argmax}_j \left\{ p_{[j]} \leq \frac{j\alpha}{m(1-p)} \right\} \qquad (3.16)$$

has an FDR equal to α. Different methods of estimating $(1 - p)$ have been considered in a number of papers (see [33, 34, 41, 73]) leading to FDR controlling rules which are more liberal than BH. Here, we construct a modified version of BH called MBH, where we obtain an estimator of p using the empirical Bayes method, which we then plug into formula (3.16).

In Figs. 3.5 and 3.6, we report the FDR and efficiency of four different multiple testing methods, where efficiency is defined as the ratio of the misclassification probability for the Bayes classifier to the misclassification probability for a given procedure. The results are based on 10000 replicates for BH, MBH, and parametric empirical Bayes and 3000 replicates for the full Bayes method.

Fig. 3.5 FDR as a function of p; $m = 200$, $\sigma = 1$, $\tau = \sqrt{2 \log 200} \approx 3.26$

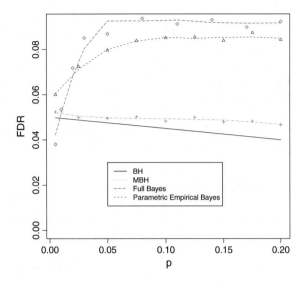

Fig. 3.6 The efficiency of multiple testing procedures as a function of p for $m = 200$, $\sigma = 1$, $\tau = \sqrt{2 \log 200} \approx 3.26$. Here, efficiency is defined as the ratio of the misclassification probability for the Bayes classifier to the misclassification probability of a given procedure

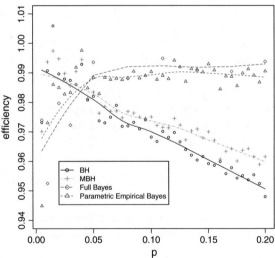

Figure 3.5 illustrates that under sparsity ($p \leq 0.2$) the modified version of BH indeed controls the FDR at the assumed level. According to Fig. 3.6, it also has a slightly larger efficiency than the original BH, particularly for larger values of p. While BH and MBH have quite similar behavior, the two Bayesian methods differ drastically. For very small p, Bayesian methods perform slightly worse than both versions of BH, which is due to the fact that for such values of p the Bayesian models struggle to identify the distribution under the alternative. The resulting estimation errors have been discussed in more detail in [12]. On the other hand, Bayes methods perform substantially better than BH for $p > 0.04$.

According to [12], all these methods have substantially lower efficiency in the case where σ is unknown. However, comparing these four methods yields qualitatively the same results as for known σ. In [39], the empirical Bayes approach to multiple testing was also shown to have very good properties within the context of minimizing the misclassification error for moderate values of sparsity. Here, the empirical Bayes procedure used a heavy-tailed Laplace prior on μ_i, as originally proposed in [52, 53], where the asymptotic minimax properties of this procedure with respect to estimating the vector of means μ were proven. The theoretical properties of this method in the context of minimizing the misclassification probability still remain to be derived.

3.2.4.6   Nonparametric Methods

In [35] a more general two-components mixture model was proposed, where the standardized test statistics are modeled as

$$T_j \sim (1 - p)F_0(x) + pF_A(x).$$

Here, F_0 and F_A are the distributions of the test statistics under the null and the alternative hypothesis, respectively.

Usually, it is assumed that F_0 is a standard normal distribution, while $F_A(x)$ is modeled nonparametrically. However, in [35] it is argued that situations may occur where both the mean and the variance of the null distribution are different from 0 and 1, respectively, and various methods are proposed for estimating these parameters, as well as the proportion of true null hypotheses, $1 - p$.

The Bayesian FDR (BFDR) and the local FDR (fdr) are closely related to the FDR and defined as follows (see [33, 34]):

$$BFDR = P(H_{0i}|H_{0i}\text{ is rejected}) = \frac{(1 - p)F_0(C)}{F(C)},$$

where $F_0(C)$ and $F(C)$ are the probabilities that the test statistic falls into the rejection region under the null hypothesis and the marginal mixture distribution, respectively, and

$$fdr(z) = \frac{(1 - p)f_0(z)}{f(z)},$$

where f_0 and $f(z)$ are the densities of the null and the mixture distribution.

To estimate $BFDR$ or $fdr(z)$, one needs to estimate p, as well as the null and the mixture densities. Many different methods are available for this. We refer the interested reader to [21, 34, 35, 51, 73] or [17].

Some empirical results illustrating the performance of nonparametric Bayes methods for multiple testing are reported in [12]. These methods adopt nonparametric estimates of the distribution of μ_i under the alternative. Interestingly, under sparsity and a relatively small m, $m = 200$, these methods performed substantially worse than the parametric Bayes methods based on a normal prior on μ_i, even when the assumption of normality was violated. These results illustrate that there seem to be major problems with estimating the mixture components in the case when p is very small. Further information about the framework and properties of nonparametric Bayes procedures for multiple testing can be found in [27] or [60], while the general statistical framework of Bayesian multiple testing in the context of decision theory is discussed in [61].

3.3 Model Selection

The main emphasis of this book is on the application of certain procedures for model selection in QTL mapping and genetic association studies. A good review on different model selection procedures is [64]. As we will see, to some extent there is a close relationship between multiple testing and model selection, but the paradigm underlying model selection is quite different from that of statistical testing. Here, the aim is not to test whether a given model is correct or not. Instead, one has a set of candidate models, and model selection provides the means to compare how well different models fit the data.

In the 1970s, Akaike developed an information theoretic approach to model selection, which is thoroughly presented in [20]. Making use of the theory of maximum likelihood and minimizing the expected Kullback–Leibler information, Akaike [2, 3] derived an information criterion (AIC) which is based on an estimate of the relative distance between a given model and 'reality'. A few years later, Schwarz [65] derived the Bayesian information criterion (BIC), which from the perspective of Bayesian model selection is based on an approximation of the posterior probability of a given model. We discuss these classical criteria in more detail in Sects. 3.3.2 and 3.3.3.

Both AIC and BIC were developed for the situation where the sample size n is substantially larger than the number of candidate models. In [20] there is an explicit warning against 'data dredging' and the use of these approaches to model selection when the number of possible models is huge. However, in the particular genetic applications described in this book, experiments are performed that inevitably lead to data sets that are exactly of this form. In the case of m (binary) genetic markers, there are 2^m possible combinations of markers to base linear models upon. This is an astronomic number even for moderate m. Even worse, geneticists are particularly interested in epistatic effects, which are typically modeled by interaction effects. There are $\binom{m}{2}$ pairwise interactions, which gives $2^{m+\binom{m}{2}}$ candidate models when both additive effects and pairwise interactions are taken into consideration.

At first glance, it seems impossible to consider model selection in such a situation, both from a practical and a statistical point of view. The first barrier is that the number of potential models is so huge that it is simply not feasible to assess all of them according to a given criterion. Furthermore, it is usually not even possible to fit statistical models of size $m > n$. The notion of sparsity, which was introduced in Sect. 3.2.4, is particularly useful in this context. In genetic association studies, only a relatively small number of markers are expected to have a large influence on the trait in question. Therefore, only models of size $k \ll n$ are considered. This still leads to a rather large number of candidate models, but methods for model selection specifically designed for this particular situation are presented in Sect. 3.3.4. Finally, in Sect. 3.3.5 we summarize various alternative approaches to model selection, which are also very useful for analyzing high-dimensional data.

3.3.1 The Likelihood Function

Akaike introduced his information theoretical approach to model selection as an extension to the theory of maximum likelihood. Therefore, the first step toward formalizing his ideas is to introduce the concept of likelihood, which is due to Fisher. We reduce our discussion here to parametric models and start with continuous random variables. Assume that measurements for a sample of size n are distributed according to the independent random variables $Y_1, \ldots, Y_n$, which all have a common density function f_θ. The parameter vector $\theta \in \Omega_\theta$, where Ω_θ is assumed to be an open subset of $\mathbb{R}^K$.

The likelihood function:
For a random sample of size n the likelihood is defined as

$$\mathscr{L}(\theta | Y_1, \ldots, Y_n) := f(Y_1, \ldots, Y_n | \theta) = \prod_{i=1}^{n} f_\theta(Y_i). \tag{3.17}$$

For continuous random variables, the likelihood function is simply the joint density of the vector $Y_1, \ldots, Y_n$, but interpreted as a function of the parameter vector for fixed data. In the case of discrete random variables (see Sect. 6.5), the probability mass function takes the role of the density function. Using maximum likelihood (ML) estimation, one looks for the parameter $\hat{\theta}$ which maximizes $\mathscr{L}(\theta | Y_1, \ldots, Y_n)$ for the data given. Often the log-likelihood, $\log \mathscr{L}$ is considered for further analysis.

In practice, the real distribution of the data (which might be characterized by density function f^*) will be unknown. Often, one would like to consider a variety of different models to fit the data. A parametric model is specified by f_θ, the density function (for continuous random variables) or the probability mass function (in the discrete case). Thus the term model actually denotes a whole family of density functions. For each model, the ML estimate yields one possibility of fitting a model to the data given. The task of model selection is to consider a set of candidate models and decide which one of these fits the data best.

Let us consider some standard examples of likelihood functions in the context of genetic association. Throughout this book we are concerned with the relationship between a phenotype Y and the genotypes of m markers $X_1, \ldots, X_m$. Often, the genotype data from each individual are considered as known fixed effects. If the phenotype Y is a quantitative trait, then multiple linear regression can be used to model genetic association, as extensively discussed in Chaps. 4 and 5 (see also [67]). A basic introduction to general linear models (GLMs) is given in Sect. 8.4.

Let $S_r \subset \{1, \ldots, m\}$ be the index vector that characterizes a particular set of $k_r = |S_r|$ genetic markers. The corresponding multiple regression model $\mathcal{M}_r$ is specified by

$$Y_i = \beta_0 + \sum_{j \in S_r} \beta_j X_{ij} + \epsilon_i \, , \quad \epsilon_i \sim \mathcal{N}(0, \sigma^2) \text{ i.i.d.} \tag{3.18}$$

Using the notation of the general likelihood setting from above, one has a parameter vector $\theta = (\beta_0, \beta_{S_r}, \sigma)$, where β_{S_r} is the k_r dimensional vector of coefficients corresponding to the markers included in the model $\mathcal{M}_r$. Therefore, the parameter space is $\Omega_\theta = \mathbb{R}^{k_r+1} \times \mathbb{R}^+$ and $K = k_r + 2$. There exist 2^m models in which the intercept β_0 is included.

The assumption that the error terms ϵ_i are independent and identically normally distributed constitutes the simplest situation for multiple regression models (see also Sect. 8.3.5). The corresponding likelihood function is given by

$$\mathcal{L}(\theta|Y_1, \ldots, Y_n) = \frac{1}{(\sqrt{2\pi}\sigma)^n} \exp\left(-\frac{RSS_r}{2\sigma^2}\right), \tag{3.19}$$

where

$$RSS_r = \sum_{i=1}^{n} \left(Y_i - \beta_0 - \sum_{j \in S_r} \beta_j X_{ij}\right)^2 \tag{3.20}$$

is the residual sum of squares corresponding to $\mathcal{M}_r$. Under the normality assumptions, the ML estimates of the parameters β_j coincide with the classical least squares estimates from Sect. 8.3.5.

A second common situation is the application of logistic regression models to a dichotomous phenotype Y. In this case, the $Y_i \sim \mathcal{B}(1, \pi_i)$ are modeled as independent Bernoulli random variables with success probability $P(Y_i = 1) = \pi_i$ such that

$$\log\left(\frac{\pi_i}{1 - \pi_i}\right) = \beta_0 + \sum_{j \in S_r} \beta_j X_{ij}. \tag{3.21}$$

This is a special case of a generalized linear model (gLM), for which a more in-depth treatment is provided in Sect. 8.5. The parameter vector in logistic regression is simply $\theta = (\beta_0, \beta_{S_r})$, and the likelihood function becomes

$$\mathcal{L}(\theta|Y_1, \ldots, Y_n) = \prod_{i=1}^{n} \left(\frac{O_{i,r}}{1 + O_{i,r}}\right)^{Y_i} \left(\frac{1}{1 + O_{i,r}}\right)^{1-Y_i}, \tag{3.22}$$

where

$$O_{i,r} = \exp\left(\beta_0 + \sum_{j \in S_r} \beta_j X_{ij}\right) \tag{3.23}$$

are the odds of success for individual i under model $\mathcal{M}_r$. There is no closed-form solution of the ML estimator for logistic regression, and one has to use numerical optimization algorithms.

If there are only two nested models to be compared, one might consider a likelihood ratio test.

The likelihood ratio test (LRT):
Let $L = \sup_\theta \mathcal{L}(\theta|Y_1, \ldots, Y_n)$ be the maximum likelihood of the larger model, and L_0 the maximum likelihood of a restricted model. Then the statistic $-2\log \frac{L_0}{L}$ is referred to as the likelihood ratio test statistic.

Under rather general assumptions (see [57]), for large n it follows that $-2\log \frac{L_0}{L}$ has approximately a chi-square distribution with ν degrees of freedom, where ν is the difference between the numbers of parameters in the two models. An LRT can be used to decide whether the restricted model explains significantly less of the variance in the data than the unrestricted model, in which case one might prefer the unrestricted model. However, if there are a large number of models to be compared, then the use of the LRT becomes impractical, particularly because the various models may no longer be nested. For a detailed discussion of the problems involved in using the LRT for model selection see [20].

Another likelihood based test that is somewhat closely related to the LRT is the **score test**. In the case of a one-parameter model ($\theta \in \mathbb{R}$), the score function is defined as the derivative of the log-likelihood,

$$S(\theta) = \frac{d}{d\theta} \log \mathcal{L}(\theta|Y_1, \ldots, Y_n),$$

and the Fisher information is defined as

$$I(\theta) = -E\left(\frac{d^2}{d\theta^2} \log \mathcal{L}(\theta|Y_1, \ldots, Y_n)\right).$$

Under some mild regularity assumptions on the underlying parametric model (see [57] again for details), one can show that the score function is asymptotically normally distributed with variance equal to the Fisher information, $S(\theta) \overset{a}{\sim} \mathcal{N}(0, I(\theta))$. This is the theory behind score tests, which can serve as an alternative to the LRT. An excellent review on score tests can be found in [7], where generalization to the case of more than one parameter is also discussed.

We consider conditional score tests in Sects. 5.3.3 and 5.3.4, where they are used as part of a more complex strategy of model selection. Specifically, in the context of logistic regression, the question arises as to whether a given model $\mathcal{M}_r$ should be extended by including some additional regressors. For example, one might perform an LRT to preselect a list of candidates. However, the score test has the big advantage that it only requires the ML estimator under the null hypothesis, which is simply the estimate under the model $\mathcal{M}_r$. If a large number of additional markers are available (which is certainly the case in GWAS), then preselection based on the score test is computationally much less intensive than using the LRT.

3.3.2 Information Theoretical Approach

The key notion of Akaike's approach to model selection is the specification of a function describing the distance between different models. In the case of a continuous variable, the distribution function of Y might be characterized by a density function f^*. While this 'reality' is unknown, one can work with f^* at a conceptual level. A good model should have density function f_θ which is, in some sense, close to f^*. Akaike used the concept of Kullback–Leibler Information [55] to measure the distance between f^* and f_θ.

Kullback-Leibler (KL) Information:

$$I(f^*, f_\theta) := \int_{x \in \Omega_x} f^*(x) \log \left(\frac{f^*(x)}{f_\theta(x)} \right) dx. \tag{3.24}$$

In the case of discrete random variables, one has

$$I(f^*, f_\theta) := \sum_{x \in \Omega_x} f^*(x) \log \left(\frac{f^*(x)}{f_\theta(x)} \right),$$

where f^* is the unknown 'true' probability mass function and f_θ is the probability mass function for a parametric model.

In information theory, the KL measure of information is also referred to as relative entropy. It measures the information which is lost when f^* is approximated by f_θ. The KL measure of information is not symmetric, but can be interpreted as a directed distance measure, because $I(f^*, f_\theta) \geq 0$ and equality only holds if $f^* = f_\theta$. Given a parametric model f_θ, one might like to find the parameter θ_0 which minimizes $I(f^*, f_\theta)$. In practice, this is not possible, as f^* is unknown. Instead, Akaike [3] showed that it is possible, up to a constant, to estimate the expected KL information $E_Y[I(f^*, f_{\hat{\theta}})]$, where $\hat{\theta} = \hat{\theta}(Y_1, \ldots, Y_n)$ depends on the data. Under certain technical conditions, it follows that

$$E_Y[I(f^*, f_{\hat\theta})] \approx -\log \mathscr{L}(\hat\theta|Y_1, \dots, Y_n) + K + \text{const},$$

where, as before, K is the dimension of the parameter vector θ. A detailed account of the derivation of this expression is given in [20]. This is the basis for Akaike's criterion for model selection.

Akaike's Information Criterion (AIC):

$$\text{AIC} := -2\log \mathscr{L}(\hat\theta|Y_1, \dots, Y_n) + 2K. \tag{3.25}$$

According to Akaike's Information Criterion, one selects the model for which the AIC score is minimized. The task of model selection is thus accomplished by solving an optimization problem: Find the minimum of (3.25) among all candidate models. AIC was motivated purely by information theoretic principles, but it is useful to note that it belongs to the category of penalized likelihood criteria. Such objective functions comprise a log-likelihood term with a negative sign, and a penalty term, here $2K$, with a positive sign. Typically a model with more parameters will enable a better fit to the data, leading to a larger log-likelihood. On the other hand, more complex models give rise to a larger penalty. So there is a tradeoff between fitting the data and parsimony.

The exact value of the AIC score is irrelevant to the problem of model selection, only the differences between the AIC scores for the candidate models are important. Selection based upon Akaike's Information Criterion suggests that the model for which the AIC score is minimized gives the best fit to the data. However, there is more information to be gained from this approach. In many applications, it is not the case that there exists one model which explains the data perfectly and all other models are wrong. Often, there are a number of models that fit the data rather well.

Assume that there is a set of R candidate models $\mathscr{M}_r, r \in \{1, \dots, R\}$, with corresponding AIC scores $AIC(\mathscr{M}_r)$. The differences between the $AIC(\mathscr{M}_r)$ can be used to gain information about the quality of each model. Let $\Delta_r = AIC(\mathscr{M}_r) - \min_r AIC(\mathscr{M}_r)$. Obviously, for the best model according to this criterion $\Delta_r = 0$, but small values of Δ_r for other models indicate that they also fit the data well. An in-depth discussion is given in [20], where relative weights based on AIC scores are introduced to provide a notion of the probability of each model. We postpone this discussion until the next section, as this type of consideration appears to be more natural in a Bayesian context.

Apart from the original AIC, there exist a number of related information criteria. In particular, we should mention the criterion

$$AIC_c = AIC + \frac{2K(K+1)}{(n-K-1)},$$

which is recommended for small samples. The additional penalty term stems from a second-order correction for small n. Another important version is $QAIC$, which was derived to deal with the problem of overdispersion in count data. For details, we again refer to the reader to [20].

3.3.3 Bayesian Model Selection and the Bayesian Information Criterion

For the sake of clarity, in this section we will slightly adapt the notation to emphasize the dependence of density functions on a particular model. Consider the random vector $Y = (Y_1, \ldots, Y_n)$. Its joint density $f_\theta(Y)$ according to the model $\mathcal{M}_r$ is denoted as $f(Y|\theta_r, \mathcal{M}_r)$. From (3.24), this can be again interpreted as a function of θ_r, but in a Bayesian context we prefer the density interpretation. From a Bayesian perspective, a prior density $g(\theta_r)$ for the parameter is also needed to fully characterize the model $\mathcal{M}_r$. Furthermore, it is required that each model itself receives a prior probability $\pi(\mathcal{M}_r)$. The key to Bayesian model selection is then simply to compute the posterior probability of each model given the data.

Bayesian Model Selection:
The posterior probability of a model given the data vector Y is

$$P(\mathcal{M}_r|Y) = \frac{P(Y|\mathcal{M}_r)\pi(\mathcal{M}_r)}{\sum_l P(Y|\mathcal{M}_l)\pi(\mathcal{M}_l)}, \tag{3.26}$$

where

$$P(Y|\mathcal{M}_r) = \int f(Y|\theta_r, \mathcal{M}_r)g(\theta_r)d\theta_r. \tag{3.27}$$

In practice, the computation of the denominator on the right-hand side of (3.26) is difficult and usually involves techniques like Markov Chain Monte Carlo (MCMC). Such a fully Bayesian approach to model selection is addressed comprehensively in [23]. The two main approaches are Stochastic Search Variable Selection and Reversible Jump MCMC.

The simplest way to make use of (3.26) in model selection is to consider those models with the largest posterior probability. For such a comparison, knowledge of the denominator $P(Y)$ is not absolutely necessary, as it is constant over all the models. This is the starting point for the derivation of the Bayesian Information Criterion (BIC) [65]. Using the Laplace approximation, it can be shown that, under appropriate regularity conditions on $f(Y|\theta_r, \mathcal{M}_r)$ and $g(\theta_r)$, the following approximation of (3.27) holds for fixed model size K and large n:

$$P(Y|\mathscr{M}_r) \approx \mathscr{L}(\hat{\theta}_r|Y)n^{-K/2}(2\pi)^{K/2}g(\hat{\theta}_r)|H_n(\hat{\theta}_r)|^{-1/2},$$

where H_n is the Hessian matrix of $-\frac{1}{n}\log\mathscr{L}(\theta|Y)$. Taking logarithms, multiplying by a factor -2, ignoring terms that stay bounded as $n \to \infty$ and disregarding the prior probabilities of the model leads to the following criterion for model selection.

Schwarz's Bayesian Information Criterion (BIC):

$$\text{BIC} := -2\log\mathscr{L}(\hat{\theta}|Y) + K\log n. \tag{3.28}$$

The Bayesian information criterion suggests selecting the model for which the BIC score is minimized. Like AIC, it is a penalized likelihood criterion, but for $n > 8$ the penalty, $K\log n$, is more severe than the AIC penalty, $2K$. Therefore, BIC has the tendency to select smaller models than AIC.

The asymptotic assumptions under which BIC is derived require that $n \to \infty$, while the number of candidate models is kept fixed. As Schwarz [65] pointed out the exact choice of prior probabilities for each model is asymptotically irrelevant in this setting, which is the main reason why $\pi(\mathscr{M}_r)$ can be ignored when considering the posterior probability of a model. However, for high-dimensional data the choice of prior probabilities for each model becomes crucial, as we will see in the following section.

One of the key properties of BIC is that it is consistent, which means the following: If the correct model is among the (fixed number of) candidate models, then as $n \to \infty$, the probability that BIC selects the correct model converges to one. Although AIC is not consistent, it does have certain asymptotic optimality properties with respect to prediction.

The philosophical principles of Bayesian model selection are rather different from the information theoretic approach. In particular, [20] is written from a purely information theoretic perspective and is rather critical of Bayesian model selection. For a concise justification of the Bayesian approach, see [44], where the merits of AIC and BIC are compared from a Bayesian perspective. In particular, AIC can also be given a specifically Bayesian interpretation.

The quality of variable selection in a Bayesian framework is closely related to the choice of a loss function. For a small number of variables, BIC is a good choice in the case when a zero-one loss function is applied to the choice of whether a variable should be included in the model or not, AIC is preferable in the case of a squared error loss function on the estimated response variable. In other words, AIC is more suitable when the aim is prediction, whereas BIC is suitable when the aim is selecting the variables in a model. However, both criteria were developed for the situation where the number of candidate models is rather small, but the sample size is large. For regression models such as (8.9) and (3.21), this means that $m \ll n$. The following section will introduce modifications of BIC which are particularly designed for the situation where m is large compared to n.

3.3.4 Modifications of BIC for High-Dimensional Data Under Sparsity

Broman and Speed [18] were among the first to apply an approach to model selection to determine the association between genetic markers and quantitative traits in experimental populations. They showed that both AIC and BIC have a tendency to select overly large models in QTL mapping. In [11], this behavior was explained as follows: In the derivation of BIC (3.28), the prior probabilities of models were neglected, which is equivalent to choosing a uniform prior distribution $\pi(\mathcal{M}_r) = \frac{1}{R}$ for all models. While this choice is uninformative regarding the likelihood of each particular model, it gives prior information on other characteristics, for example on the dimension of a model (see [11]).

In a regression setting, such as (8.9) or (3.21), there are $\binom{m}{k}$ different models that include k regressors. When each model has the same prior probability, it follows that for combinatorial reasons models of size around $m/2$ are likely to be chosen, while both very small and very large models are unlikely to be chosen. In the situation of high-dimensional data under sparsity, where one expects that the true model will only include a few regressors, this has the effect that BIC will select models which tend to be too large.

To rectify this situation, Bogdan et al. [11] assume that the prior probability of a model depends on its dimension. In particular, they assume that the prior probability of a model is of the form

$$\pi(\mathcal{M}_r) = \omega^{k_r}(1 - \omega)^{m-k_r},$$

a common choice in Bayesian model selection. Here, ω can be interpreted as the a priori expected proportion of true signals, where all the regressors have the same chance of being included in the model, independently of one other. Using this prior distribution in the derivation of BIC, yields the following modified criterion (see [11, 15] for more details).

FWER controlling modification of BIC (mBIC):

$$\text{mBIC} := -2\log \mathcal{L}(\hat{\theta}|Y) + K\log\left(\frac{nm^2}{c^2}\right) = BIC + 2K\log\left(\frac{m}{c}\right).$$
$$(3.29)$$

Here $c = m\omega$ is the expected number of regressors in the true model. If there is no prior knowledge regarding this expected number, then a standard choice of $c = 4$ is suggested in [15], which guarantees control of the FWER at a level of 0.1 for $n \geq 200$ and $m \geq 10$.

As pointed out in [15], there is a strong relationship between this modification of BIC for high-dimensional data (3.29) and corrections for multiple testing. In particular, in the case of orthogonal regressors and a known error variance, the estimate of a regression coefficient for a given explanatory variable does not depend on other regressors. Therefore, in this simplistic scenario, model selection becomes equivalent to multiple testing. Under orthogonality, it turns out that selection based on mBIC is equivalent to a Bonferroni correction for multiple testing, which gives a precise meaning to the statement that mBIC controls FWER.

In [40], it was proven that, under orthogonality, mBIC is ABOS only in the case of extreme sparsity. These results were obtained both for known and unknown error variance σ^2, and they are closely related to the results on ABOS for the Bonferroni rule. This optimality follows from Sect. 3.2.4, where the FDR controlling Benjamini–Hochberg procedure is ABOS under much less restrictive assumptions on sparsity than the Bonferroni rule, motivating the search for modifications of BIC which control FDR under sparsity.

The starting point for this endeavor is the approach to model selection discussed in [1], which applies penalties of the form

$$\sum_{l=1}^{K} q_N^2(\alpha l/2m), \tag{3.30}$$

where $q_N(\eta)$ is the $(1-\eta)$—quantile of the standard normal distribution (see also [8, 37, 76] or [43] for a related approach). Under orthogonality, model selection with such penalties is closely related to BH and the corresponding step-down FDR controlling procedure (SD). In [40, 81], approximations of penalties of the form (3.30) lead to the following modifications of BIC.

FDR controlling modifications of BIC:

$$\text{mBIC1} := -2\log \mathcal{L}(\hat{\theta}|Y) + K \log\left(nm^2\right) - 2\log(K!) - \sum_{i=1}^{K} \log\log(nm^2/i^2),$$
$$\tag{3.31}$$

$$\text{mBIC2} := -2\log \mathcal{L}(\hat{\theta}|Y) + K \log\left(\frac{nm^2}{c^2}\right) - 2\log(K!) = mBIC - 2\log(K!).$$
$$\tag{3.32}$$

The first criterion, mBIC1, is derived using a second-order approximation of (3.30), whereas mBIC2 stems from a first-order approximation. In [40], it is shown that both criteria are ABOS under much softer constraints on sparsity levels than mBIC. These FDR controlling procedures adapt well to the unknown level of sparsity, a result that resonates with the findings of [1], and with the discussion on the properties of BH given in Sect. 3.2.3.

In Fig. 3.7, the results of a small simulation study are presented, which illustrates the performance of the various criteria for model selection (for more details see [40]). The first column shows the estimated FDR, power, and misclassification rate in the case of extreme sparsity. This refers to the fact that the number of true effects was kept constant at $K = 12$, while the total number of regressors, as well as the sample sizes varied. One can clearly see the weakness of BIC under sparsity. It tends to select overly large models. It thus has greater power than the other criteria, but also a much larger FDR and, as a result, a strongly inflated misclassification rate. mBIC is more conservative than both mBIC1 and mBIC2. The FDR of mBIC is close to zero and its power is lower than that of mBIC1 and mBIC2. However, in terms of the misclassification rate, these three modifications perform almost equally well for larger m, which illustrates nicely the fact that, in the case of extreme sparsity, mBIC, mBIC1, and mBIC2 are all ABOS. For small m, mBIC2 performs slightly worse than mBIC1, which is due to the fact that the first-order approximation of (3.30) is not yet sufficiently accurate.

The second column of Fig. 3.7 presents results where the sparsity level $p = K/m$ is kept fixed roughly at 0.1. Hence, we are no longer in the situation that $p \to 0$. In this situation, BIC still selects overly large models, but at least the FDR decreases as the sample size increases. However, its misclassification rate is always greater than that of mBIC1 and mBIC2. On the other hand, mBIC now performs distinctly worse than in the case of extreme sparsity. As before, its FDR remains close to 0, but compared with mBIC1 and mBIC2, the differences between the powers of these procedures are now larger than in the case of extreme sparsity. The misclassification rate of mBIC for large n and m is close to that of BIC, but is visibly larger than that of mBIC1 and mBIC2. This illustrates the adaptivity of the FDR controlling procedures to an unknown level of sparsity. For large m, mBIC1 and mBIC2 have the minimal misclassification rate within the whole range of sparsity characterized by the extreme situations of $K = $ constant (extreme sparsity) and $K/m = $ constant ≈ 0.1 (not very sparse).

3.3.5 Further Approaches to Model Selection

Over the past two decades, the LASSO (least absolute shrinkage and selection operator, [75]) has become an extremely popular method for the regression analysis of high-dimensional data. An excellent monograph dealing almost entirely with the theory and application of the lasso is [19]. The lasso procedure is defined as follows:

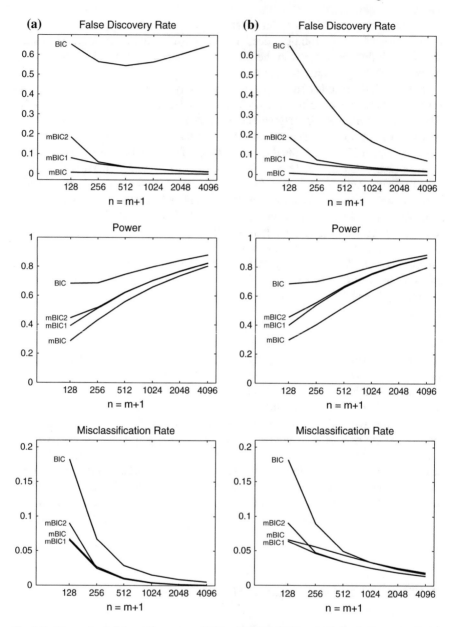

Fig. 3.7 Comparison of the performance of BIC, mBIC1, mBIC2 and mBIC at different levels of sparsity. The *first column* illustrates the case of extreme sparsity. In the *second column* the sparsity level is kept constant. The *first line* shows the estimated FDR, the *second line* the estimated power, and the *third line* the misclassification rate. **a** Extreme sparsity. **b** Constant sparsity

LASSO: Find the parameter vector θ which minimizes the penalized likelihood function

$$- \log \mathscr{L}(\theta|Y) + \lambda \sum_{j=1}^{m} |\beta_j| \, , \text{ where } \lambda > 0. \qquad (3.33)$$

Compared with the criteria from the previous section, the log-likelihood is not penalized according to the dimension of a model (l_0-norm), but rather according to the l_1-norm of the regression coefficients. This type of penalty leads to a general shrinking of the regression coefficients towards zero, such that some coefficients actually become 0. Therefore, the lasso can also be seen as a selection procedure. The amount of shrinkage and the size of the selected model will depend on the choice of the tuning parameter, λ, which has to be determined either by cross-validation procedures or by some other considerations.

Finding the minimum of (3.37) is equivalent to solving a convex quadratic program with linear inequality constraints. This optimization problem is substantially easier than minimizing information criteria, such as AIC or BIC, over all possible models. A vast number of algorithms exist for convex quadratic programming with linear constraints, see for example [16, 26]. A particularly elegant way to handle lasso computationally is discussed in [32], where a sophisticated forward selection technique called least angular regression (LAR) is introduced. The connection between LAR and lasso leads to an extremely efficient algorithm to solve (3.37) for the whole range of λ required.

Using general l_p—norms as penalty functions leads to so-called bridge regression [38]. The special case of l_2—penalties yields classical ridge regression [47], which was specifically developed to deal with correlated regressors. Ridge regression itself does not serve as a selection method, but its penalty is incorporated in the more recent elastic net [82], which uses both l_1 and l_2—penalties and thus attempts to combine the benefits of lasso and ridge regression.

Elastic net: Minimize the penalized likelihood function

$$- \log \mathscr{L}(\hat{\theta}|Y) + \lambda_1 \sum_{j=1}^{m} |\beta_j| + \lambda_2 \sum_{j=1}^{m} \beta_j^2 \, , \text{ with } \lambda_1, \lambda_2 > 0. \qquad (3.34)$$

Note that, compared with the lasso, now two tuning parameters have to be decided upon. Also, [82] suggests that the estimates of θ obtained by minimizing (3.34) should be rescaled by a factor $(1 + \lambda_2)$ to improve prediction. One potential benefit of the elastic net is its ability to select a whole group of correlated regressors, whereas the lasso would select only one representative of such a group. A version of the lasso which is able to select whole groups was introduced by [80].

Group lasso: Find the parameter vector θ which minimizes the penalized likelihood function

$$- \log \mathscr{L}(\theta|Y) + \lambda \sum_{g=1}^{G} \sqrt{k_g} ||\beta_{I_g}||_2, \qquad (3.35)$$

where $\lambda > 0$, k_g is the size of the g-th group of variables and I_g is the set of variables belonging to it, where $g \in \{1, \ldots, G\}$.

Sparse partial least squares (SPLS) regression is another recent development introduced in [24] to deal in particular with correlated regressors in high dimensions where $m > n$. Partial least squares had been developed in the 1960s to deal with correlated regressors [79]. SPLS also incorporates l_1 and l_2—penalties to enforce sparse solutions.

Currently, the whole field of model selection under sparsity is rapidly developing, and there is a lot of research going on to investigate the advantages and disadvantages of all these different methods. The approaches to model selection discussed in this section share the trait that they depend on the choice of certain parameters to define the selection criterion. In practice, one widely used strategy is to determine these parameters by cross-validation, which usually leads to good results in terms of prediction. However, when one is interested in selecting variables for a model, then cross-validation potentially leads to the selection of overly complex models. An extensive debate on these issues can be found in [32], where in particular the comments of Ishwaran stress the point that prediction accuracy is not necessarily associated with selecting the appropriate variables for a model. This discussion is closely connected with the different properties of AIC and BIC that were mentioned at the end of Sect. 3.3.3.

3.3.5.1   SLOPE

In this section, we briefly discuss Sorted L-One Penalized Estimation (SLOPE), proposed in [13, 14], which is a generalization of LASSO.

Consider the classical regression model

$$Y_i = \sum_{j=1}^{m} \beta_j X_{ij} + \epsilon_i, \quad i = 1, \ldots, n, \qquad (3.36)$$

where $\epsilon_1, \ldots, \epsilon_n$ are independent random variables from the normal $N(0, \sigma^2)$ distribution. Using the matrix notation: $Y = (Y_1, \ldots, Y_n)'$, $X = [X_{ij}]_{n \times m}$ and $\epsilon = (\epsilon_1, \ldots, \epsilon_n)'$, model (3.36) can be written in the form:

$$Y = X\beta + \epsilon.$$

It is well known that for LASSO to estimate the multiple regression model (3.36) well, the smoothing parameter λ should be proportional to σ. Therefore, in the case when σ is known, LASSO is often written in the following alternative form:

LASSO with known σ: Find the parameter vector β which minimizes the penalized residual sum of squares

$$1/2||Y - X\beta||^2 + \lambda\sigma \sum_{j=1}^{m} |\beta_j|, \text{ where } \lambda > 0. \qquad (3.37)$$

Let us now consider an orthogonal design: $X'X = I$. In this case, it can be shown that the above version of LASSO selects those explanatory variables for which $\frac{\hat{\beta}}{\sigma} > \lambda$, where $\hat{\beta} = X'Y$ is the regular least squares estimator of β. Note also that when $\beta = 0$, then $\frac{\hat{\beta}}{\sigma} \sim N(0, 1)$. Thus, to control the FWER at the level α under an orthogonal design, λ may be selected according to the Bonferroni rule, i.e., $\lambda = \Phi^{-1}\left(1 - \frac{\alpha}{2m}\right) \approx \sqrt{2 \log m}$. Using the same arguments as for mBIC1, it can be shown that under orthogonal designs the appropriate version of LASSO is ABOS under extreme sparsity, i.e., when the number of true signals remains approximately constant as $m \to \infty$.

Furthermore, in [22] it is shown that the choice of $\lambda = c\sqrt{2 \log m}$, for some $c > 1$, leads to a procedure for model selection which has very good asymptotic properties with respect to both optimizing the estimation error and the probability of identifying the correct model. In [22], the assumption of orthogonality is replaced by the coherence property

$$\sup_{1 \leq i \leq j \leq m} | < X_i, X_j > | \leq \frac{C}{\log m}, \qquad (3.38)$$

where $C > 0$. It is also required that the number of true signals K satisfies

$$K \leq \frac{c_0 m}{||X||^2 \log m}, \qquad (3.39)$$

where $c_0 > 0$ and $||X||$ denotes the operator norm of X. As argued in [22], the conditions (3.38) and (3.39) are milder than the assumptions under which the consistency and optimality of LASSO were previously proved.

Since choosing $\lambda \approx \sqrt{2 \log m}$ corresponds to the Bonferroni correction for multiple testing, it is natural to modify LASSO so that it is based on a decreasing sequence of threshold values, similar to BH. Applying this reasoning, [13, 14] introduced a new procedure for model selection, called SLOPE (Sorted L_1 Penalized Estimation).

SLOPE: Let λ be a nonincreasing sequence of nonnegative scalars,

$$\lambda_1 \geq \lambda_2 \geq \cdots \geq \lambda_m \geq 0, \tag{3.40}$$

with $\lambda_1 > 0$. SLOPE finds the parameter vector β which minimizes

$$1/2||Y - X\beta||^2 + \sum_{i=1}^{m} \sigma\lambda_i |\beta|_{(i)}, \tag{3.41}$$

where $|\beta|_{(1)} \geq |\beta|_{(2)} \geq \cdots \geq |\beta|_{(m)}$ are the order statistics of the absolute values of the coordinates of β.

When $\lambda_1 = \lambda_2 = \cdots = \lambda_m$, then SLOPE reduces to a regular LASSO.

As discussed in [14], SLOPE involves solving a convex optimization problem and is thus tractable. In [14], an efficient algorithm for solving optimization problem (3.41) is provided. The computational cost of this algorithm is roughly the same as that of solving the optimization problem resulting from standard LASSO.

If $X'X = I$, then the choice of $\lambda_i = \Phi^{-1}\left(1 - \frac{i\alpha}{m}\right)$ allows SLOPE to control the FDR at the level $\alpha \frac{m_0}{m}$. Also, it can be shown that the set of detections using the SLOPE procedure is sandwiched between the sets of detections made under the step-up and step-down versions of BH and therefore SLOPE possesses the asymptotic optimality properties of BH, described in [1, 9].

In the case when the design is non-orthogonal, the sequence λ needs to be adjusted to take into account the correlation structure between different regressors. A heuristic adjustment procedure is proposed in [13], which, according to the simulation study, performs very well when the regressors are not too strongly correlated. Figure 3.8 compares SLOPE to LASSO in the case when the elements of the design matrix are generated as independent normal $N(0, 1/\sqrt{n})$ random variables for $n = m = 5000$. The number k of true regressors varies between 1 and 50 and their effect sizes are equal to $\beta_1 = \cdots = \beta_k = \sqrt{2 \log m} \approx 4.1$. The smoothing parameter λ for LASSO is selected in two ways: by cross-validation and using a fixed Bonferroni type choice to control the FWER at the level $\alpha = 0.1$; $\lambda = \sigma\Phi^{-1}\left(1 - \frac{0.1}{2m}\right)$. Similarly, the sequence of the smoothing parameters for SLOPE is selected so as to control the FDR at the level $q = 0.1$.

Figure 3.8 illustrates that for sparse designs SLOPE maintains the FDR at almost exactly the desired level of 0.1. The behavior of LASSO with a fixed λ is quite different. Its FDR decreases from 0.1 when there are no true signals to only 0.014 when the number of true regressors, k, equals 50. This decrease in the FDR as a

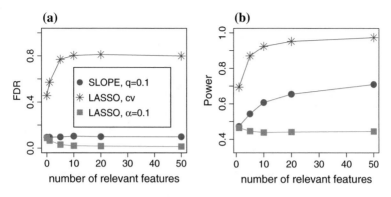

Fig. 3.8 The FDR and power of various procedures for model selection

function of k is a general property of methods that control the FWER. The price for this conservativeness is a loss in power in comparison to SLOPE. In our example, the power of LASSO is close to 44 % and remains stable over the whole range of k between 1 and 50. Instead, the power of SLOPE increases from 47 % for $k = 1$ to almost 71 % for $k = 50$. It is also interesting to observe that, in our example, cross-validation has rather disastrous properties with respect to model selection. For k between 10 and 50 its FDR equals 0.8, which means that a large majority of detections correspond to false regressors.

In the case when the variance of the error term is unknown, Bogdan et al. [13] propose an iterative version of SLOPE, which jointly estimates σ and β. This method is similar to the scaled LASSO, proposed in [74]. This procedure is based on a standardized regression model: the vectors of the responses and of the explanatory variables are centered so that they have a zero mean, and the columns of the design matrix are additionally standardized so that they have a unit l_2 norm. The procedure starts by estimating σ^2 using the variance of Y and then iteratively invokes SLOPE and estimates σ^2 using a regular unbiased estimator $\hat{\sigma}^2 = \frac{RSS}{n-l-1}$, where RSS is the classical residual sum of squares in the current model with l predictors. Since σ^2 is initially overestimated, the model selected in the first iteration is usually too small and the number of identified regressors tends to increase at each stage. The procedure stops when the next iteration does not change the currently identified model. As reported in [13], this procedure converges rather quickly and enables control of the FDR at a level close to the nominal FDR. This algorithm is implemented in the R package "SLOPE", available on the Comprehensive R Archive Network (CRAN).

Figure 3.9 compares the performance of SLOPE and of the classical marginal tests based on simulating the effect of QTLs. We set $n = m = 5000$, and simulate n genotypes of m independent Single Nucleotide Polymorphisms (SNPs). For each of these SNPs, the minor allele frequency (MAF) is generated from the uniform distribution on the interval $(0.1, 0.5)$. The explanatory variables are defined for each possible genotype as follows:

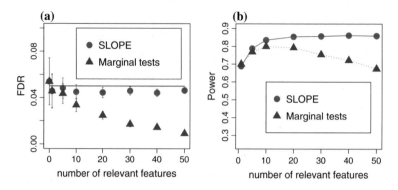

Fig. 3.9 The FDR and power for SLOPE and marginal tests

$$\tilde{x}_{ij} = \begin{cases} -1 \text{ for } aa \\ 0 \text{ for } aA \\ 1 \text{ for } AA \end{cases}, \qquad (3.42)$$

where a and A denote the minor and reference alleles at the j-th SNP for the i-th individual. The matrix $\tilde{X}$ is centered and standardized, so that the columns of the final design matrix X have zero mean and unit norm. The trait values are simulated according to the model

$$y = X\beta + z, \qquad (3.43)$$

where $z \sim N(0, I)$. Hence, we assume that only additive effects exist and there is no interaction between loci (epistasis). We vary the number of nonzero regression coefficients, k, between 0 and 50 and set their size to $1.2\sqrt{2\log m} \approx 4.95$ (a 'moderate' signal). For each value of k, 500 replicates are performed, with the set of k factors being chosen at random each time. Since our design matrix is centered and does not contain an intercept, we also center the vector of responses and treat $\tilde{y} = y - \bar{y}$ as our observations of the trait of interest, where $\bar{y}$ is the mean of y. We compare the SLOPE procedure controlling the FDR at a level of 0.05 with classical single marker (marginal) tests, based on simple regression models, which include only one SNP at a time. The p values from marginal tests are adjusted using the Benjamini–Hochberg procedure at the FDR level $q = 0.05$.

According to Fig. 3.9, the SLOPE procedure adapted to Gaussian designs works very well when the design matrix contains the genotypes of independent SNPs. The empirical FDR is controlled at a level of 0.05 and for large k the power of SLOPE substantially exceeds the power of single marker tests. The loss of power for single marker tests is due to the fact that the true regressors inflate the background residual error, which is not accounted for by simple regression models. This leads both to a loss in power and a fall in the empirical FDR, which for large k is substantially below the assumed level. Interestingly, the simulations reported in [13] also suggest that SLOPE is more robust to deviations from the assumptions of the model (such as nonlinearity or non-normality of the error terms) than single marker tests.

In the case of real GWAS studies, where neighboring SNPs are strongly correlated, control of the FDR is still possible using appropriate adaptations of SLOPE, which involve grouping correlated SNPs. The related package *geneSLOPE* for GWAS for quantitative traits is available on github. Updates including case-control studies and robust versions of SLOPE will follow soon.

References

1. Abramovich, F., Benjamini, Y., Donoho, D.L., Johnstone, I.M.: Adapting to unknown sparsity by controlling the false discovery rate. Ann. Stat. **34**, 584–653 (2006)
2. Akaike, H.: A new look at the statistical model identification. IEEE Trans. Autom. Control **19**(6), 716–723 (1974)
3. Akaike, H.: Information theory and an extension of the maximum likelihood principle. In: Proceedings of the 2nd International Symposium on Information Theory, 267–281 (1973)
4. Benjamini, Y., Hochberg, Y.: Controlling the false discovery rate: a practical and powerful approach to multiple testing. J. R. Stat. Soc. B **57**, 289–300 (1995)
5. Benjamini, Y., Hochberg, Y.: On the adaptive control of the false discovery fate in multiple testing with independent statistics. J. Educ. Behav. Stat. **25**, 60–83 (2000)
6. Benjamini, Y., Yekutieli, D.: The control of the false discovery rate in multiple testing under dependency. Ann. Stat. **29**(4), 1165–1188 (2001)
7. Bera, A.K., Bilias, Y.: Rao's score, Neyman's $C(\alpha)$ and Silvey's LM tests: an essay on historical developments and some new results. J. Stat. Plan. Infer. **97**, 9–44 (2001)
8. Birgé, L., Massart, P.: Gaussian model selection. J. Eur. Math. Soc. (JEMS) **3**, 203–268 (2001)
9. Bogdan, M., Chakrabarti, A., Frommlet, F., Ghosh, J.K.: Asymptotic Bayes-optimality under sparsity of some multiple testing procedures. Ann. Stat. **39**, 1551–1579 (2011)
10. Bogdan, M., Frommlet, F., Szulc, P., Tang H.: Model selection approach for genome wide association studies in admixed populations. Technical Report (2013)
11. Bogdan, M., Ghosh, J.K., Doerge, R.W.: Modifying the Schwarz Bayesian information criterion to locate multiple interacting quantitive trait loci. Genetics **167**, 989–999 (2004)
12. Bogdan, M., Ghosh, J.K., Tokdar S.T.: A comparison of the Simes-Benjamini-Hochberg procedure with some Bayesian rules for multiple testing. In: Balakrishnan, N., Peña, E., Silvapulle, M.J. (eds.) Beyond Parametrics in Interdisciplinary Research: Fetschrift in Honor of Professor Pranab K. Sen, IMS collections, vol. 1, pp. 211–230. Beachwood Ohio (2008)
13. Bogdan, M., van den Berg, E., Sabatti, C., Su, W., Candès, E.J.: SLOPE—Adaptive Variable Selection via Convex Optimization. Ann. Appl. Stat. **9**, 1103–1140 (2015)
14. Bogdan, M., van den Berg, E., Su, W., Candès, E.J.: Statistical estimation and testing via the sorted ℓ_1 norm. arXiv:1310.1969 (2013)
15. Bogdan, M., Żak-Szatkowska, M., Ghosh, J.K.: Selecting explanatory variables with the modified version of Bayesian Information criterion. Qual. Reliab. Eng. Int. **24**, 627–641 (2008)
16. Boyd, S., Vandenberghe, L.: Convex Optimization. Kluwer, Cambridge University Press (2004)
17. Broberg, P.: A comparative review of estimates of the proportion unchanged genes and the false discovery rate. BMC Bioinform. **6**, 199 (2005)
18. Broman, K.W., Speed, T.P.: A model selection approach for the identification of quantitative trait loci in experimental crosses. J. Roy. Stat. Soc.: Ser. B (Stat. Meth.) **64**(4), 641–656 (2002)
19. Bühlmann, P., van de Geer, S.: Statistics for High-Dimensional Data. Springer, Heidelberg (2011)

20. Burnham, K.P., Anderson, D.R.: Model Selection and Multimodel Inference, 2nd edn. Springer, New York (2002)
21. Cai, T., Jin, J.: Optimal rates of convergence for estimating the null and proportion of non-null effects in large-scale multiple testing. Ann. Stat. **38**, 100–145 (2010)
22. Candès, E.J., Plan, Y.: Near-ideal model selection by l1 minimization. Ann. Stat. **37**, 2145–2177 (2007)
23. Chipman, H., George, E.I., McCulloch, R.E.: The practical implementation of bayesian model selection. In: Lahiri, P. (ed.) Model Selection (IMS Lecture Notes), pp. 65–116. Beachwood, OH (2001)
24. Chun, H., Keles, S.: Sparse partial least squares regression for simultaneous dimension reduction and variable selection. J. Roy. Stat. Soc.: Ser. B (Stat. Meth.) **72**(1), 3–25 (2010)
25. Churchill, G.A., Doerge, R.W. Empirical threshold values for quantitative trait mapping. Genetics **138**, 963–971 (1994)
26. De Leeuw, J., Hornik, K., Mair, P.: Isotone optimization in R: Pool-Adjacent-Violators Algorithm (PAVA) and active set methods. Journal of statistical software **32** (5): 1–24, (2009)
27. Do, K., Müller, P., Tang, F.: A Bayesian mixture model for differential gene expression. Appl. Stat. **54**, 627–644 (2005)
28. Doerge, R.W., Churchill, G.A.: Permutation tests for multiple loci affecting a quantitative character. Genetics **142**, 285–294 (1996)
29. Donoho, D., Tanner, J.: Observed universality of phase transitions in high-dimensional geometry, with implications for modern data analysis and signal processing. Phil. Trans. R. Soc. A **367**, 4273–4293 (2009)
30. Dudoit, S., Shaffer, J.P., Boldrick, J.C.: Multiple hypothesis testing in microarray experiments. Stat. Sci. **18**, 71–103 (2003)
31. Dudoit, S., van der Laan, M.J.: Multiple Testing Procedures with Applications to Genomics. Springer, New York (2008)
32. Efron, B., Hastie, T., Johnstone, I., Tibshirani, R.: Least angle regression. Ann. Stat. **32**(2), 407–499 (2004)
33. Efron, B., Tibshirani, R., Storey, J.D., Tusher, V.: Empirical Bayes analysis of a microarray experiment. J. Am. Stat. Assoc. **96**, 1151–1160 (2001)
34. Efron, B., Tibshirani, R.: Empirical Bayes methods and false discovery rates for microarrays. Genet. Epidemiol. **23**, 70–86 (2002)
35. Efron, B.: Microarrays, empirical Bayes and the two-group model. Stat. Sci. **23**(1), 1–22 (2008)
36. Ferreira, J.A., Zwinderman, A.H.: On the Benjamini-Hochberg method. Ann. Stat. **34**(4), 1827–1849 (2006)
37. Foster, D.P., Stine, R.A.: Local asymptotic coding and the minimum description length. IEEE Trans. Inf. Theor. **45**, 1289–1293 (1999)
38. Frank, I.E., Friedman, J.H.: A statistical view of some chemometrics regression tools. Technometrics **35**, 109–148 (1993)
39. Frommlet, F., Bogdan, M: Some optimality properties of FDR controlling rules under sparsity. Technical Report (2012)
40. Frommlet, F., Chakrabarti, A., Murawska, M., Bogdan, M.: Asymptotic Bayes optimality under sparsity for generally distributed effect sizes under the alternative. arXiv:1005.4753 (2011)
41. Genovese, C., Wasserman, L.: A stochastic process approach to false discovery control. Ann. Stat. **32**, 1035–1061 (2004)
42. Genovese, C., Wasserman, L.: Operating characteristics and extensions of the false discovery rate procedure. J. Roy. Stat. Soc. Ser. B **64**, 499–517 (2002)
43. George, E.I. Foster, D.F.: Calibration and empirical Bayes variable selection. Biometrika **87**, 731–747 (2000)
44. Ghosh, J.K., Samanta, T.: Model selection—an overview. Curr. Sci. **80**, 1135–1144 (2001)
45. Hochberg, Y., Tamhane, A.C.: Multiple Comparison Procedures. Wiley, New York (1987)
46. Hochberg, Y.: A sharper Bonferroni procedure for multiple tests of significance. Biometrika **75**, 800–803 (1988)

47. Hoerl A.E., Kennard, R.W.: Ridge regression: biased estimation for nonorthogonal problems. Technometrics **12**, 55–67 (1970)
48. Holm, S.: A simple sequentially rejective Bonferroni test procedure. Scand. J. Stat. **6**, 65–70 (1979)
49. Hsu, J.C.: Multiple Comparisons: Theory and Methods. Chapman and Hall, New York (1996)
50. James, W., Stein, C.: Estimation with quadratic loss, Proc. Fourth Berkeley Symp. Math. Stat. Prob. **1**, 361–79 (1961)
51. Jin, J., Cai, T.C.: Estimating the null and the proportion of non-null effects in large-scale multiple comparisons. J. Am. Stat. Assoc. **102**, 495–506 (2007)
52. Johnstone, I.M., Silverman, B.W.: EbayesThresh: R programs for empirical Bayes thresholding. J. Stat. Softw. **12**(8) (2005)
53. Johnstone, I.M., Silverman, B.W.: Needles and straw in haystacks: empirical Bayes estimates of possibly sparse sequences. Ann. Stat. **32**, 1594–1649 (2004)
54. Korn, E.L., Troendleb, J.F., McShanea, L.M., Simona, R.: Controlling the number of false discoveries: application to high-dimensional genomic data. J. Stat. Plan. Infer. **124**(2), 379–398 (2004)
55. Kullback, S.: Information Theory and Statistics. John Wiley and Sons, New York (1959)
56. Lehmann, E.L., Romano, J.P.: Generalizations of the familywise error rate. Ann.Stat. **33**, 1138–1154 (2005)
57. Lehmann, E.L., Romano, J.P.: Testing Statistical Hypotheses. Springer, New York (2005)
58. Lehmann, E.L. D'Abrera, H.J.M.: Nonparametrics: Statistical Methods Based on Ranks. McGraw-Hill, New York (1975)
59. Marcus, R., Peritz, E., Gabriel, K.R.: On closed testing procedures with special reference to ordered analysis of variance. Biometrika **63**, 655–660 (1976)
60. Martin, R., Tokdar, S.T.: A nonparametric empirical Bayes framework for large-scale multiple testing. Biostatistics. **13**, 427–439 (2012)
61. Müller, P., Giovanni, P., Rice, K.: FDR and Bayesian multiple comparisons rules. In: Proceedings of the Valencia/ISBA 8th World Meeting on Bayesian Statistics. Oxford University Press (2007)
62. Neuvial, P., Roquain, E.: On false discovery rate thresholding for classification under sparsity. Ann. Stat. **40**, 2572–2600 (2012)
63. Neyman, J., Pearson, E.: On the problem of the most efficient tests of statistical hypotheses. Phil. Trans. R. Soc. Ser. A **231**: 289–337 (1933)
64. Rao, C.R., Wu, Y.: On model selection. In: Lahiri, P. (ed.) Model selection (IMS Lecture Notes), pp. 1–57. Beachwood, OH (2001)
65. Schwarz, G: Estimating the dimension of a model. Ann. Stat. **6**(2), 461–464 (1978)
66. Scott, J.G., Berger, J.O.: An exploration of aspects of Bayesian multiple testing. J. Stat. Plan. Inf. **136**, 2144–2162 (2006)
67. Seber, A.F., Lee, A.J.: Linear Regression Analysis. John Wiley and Sons (2003)
68. Seeger, P.: A note on a method for the analysis of significance en masse. Technometrics. **10**, 586–593 (1968)
69. Shaffer, J.P.: Multiple hypothesis testing. Annu. Rev. Psychol. **46**, 561–584 (1995)
70. Simes, R.J.: An improved Bonferroni procedure for multiple tests of significance. Biometrika **73**(3), 751–754 (1986)
71. Stein, C.: Inadmissibility of the usual estimator for the mean of a multivariate distribution. Proc. Third Berkeley Symp. Math. Stat. Prob. **1**, 197–06 (1956)
72. Storey, J.D.: The positive false discovery rate: a Bayesian interpretation and the q-value. Ann. Stat. **31**(6), 2013–2035 (2003)
73. Storey, J.D.: A direct approach to false discovery rates. J. R. Stat. Soc. Ser. B **64**, 479–498 (2002)
74. Sun, T., Zhang, C.-H.: Scaled sparse linear regression. Biometrika **99**(4), 879–898 (2012)
75. Tibshirani, R.: Regression shrinkage and selection via the lasso. J. Roy. Stat. Soc B. **58**(1), 267–288 (1996)

76. Tibshirani, R. Knight, K.: The covariance inflation criterion for adaptive model selection, J. Roy. Stat. Soc. B **55**, 757–796 (1999)
77. Westfall, P.H., Young, S.S.: Resampling-Based Multiple Testing. Wiley, New York (1993)
78. Wettenhall, J. M., Smyth G. K.: limmaGUI: a graphical user interface for linear modeling of microarray data. Bioinformatics **20**(18): 3705–3706 (2004)
79. Wold, H.: Estimation of principal components and related models by iterative least squares. In Krishnaiaah, P.R. (ed.) Multivariate Analysis, pp. 391–420. Academic Press, New York (1966)
80. Yuan, M., Lin, Y. Model selection and estimation in regression with grouped variables. J. Roy. Stat. Soc. Ser. B **68**(1), 49–67 (2007)
81. Żak-Szatkowska, M., Bogdan, M.: Modified versions of Bayesian information criterion for sparse generalized linear models. CSDA **55**, 2908–2924 (2011)
82. Zou, H., Hastie, T.: Regularization and variable selection via the elastic net. J. Roy. Stat. Soc B **67**(2), 301–320 (2005)

Chapter 4
Statistical Methods of QTL Mapping for Experimental Populations

4.1 Classical Approaches

4.1.1 Single Marker Tests

The first successful attempts at localizing genes influencing quantitative traits (Quantitative Trait Loci, QTL) were performed long before the development of molecular genetic markers. It was believed at the beginning of the twentieth century that quantitative traits may depend on the joint influence of many genes (see e.g., [23, 24, 32]). In his seminal article [27], Fisher introduced a genetic model of the inheritance of quantitative traits. Fisher's model included multiple genes with additive and dominance effects. He also mentioned the possibility of gene–gene interactions (epistasis). In addition, this article is fundamental to statistics as a discipline, since it contains the definition of **variance**.

The first attempt at localizing QTL can be dated to Sax [46], where classical t-tests led to the detection of an association between the weight of bean seeds and their color. This observation implies that genes responsible for bean weight are linked to genes influencing color. In this context, genes influencing bean color can be viewed as genetic markers. The study reported in [46] also suggests that the weight of bean seeds results from the cumulative effect of several genes on different chromosomes.

In subsequent years, statistical models, methods of estimating genetic parameters, and statistical tests for the association between a trait and individual genetic markers were developed (see e.g., [2, 51–54]). Using the ideas presented in the seminal paper of Soller et al. [50], we will present theoretical calculations for the power of a test based on a single marker and a backcross population. These calculations, which preceded the development of molecular markers, formed the basis for future decisions concerning the density of markers and the interpretation of tests based on single marker.

© Springer-Verlag London 2016
F. Frommlet et al., *Phenotypes and Genotypes*, Computational Biology 18,
DOI 10.1007/978-1-4471-5310-8_4

4.1.2   Power of a Test Based on a Single Marker as a Function of the Distance Between the Marker and a QTL

Let us consider the situation where one QTL influences a trait in a backcross population. Let μ_2 denote the expected value of the trait for an individual who is homozygous
at the marker (M = AA) and μ_1 denote the expected value of the trait for an individual who is heterozygous at the marker (M = aA). Moreover, we will assume that
the variance of the trait is the same in both subpopulations and denote this variance
by σ^2.

4.1.2.1   When the Marker is at a QTL

Assume that we are testing for the presence of a QTL using a marker whose location coincides with a QTL. For this purpose, we divide the pool of individuals from
the backcross population into two groups: one group contains individuals who are
homozygous at the marker, and the other one contains individuals who are heterozygous there. For simplicity, we will assume that the number of individuals in both
groups is the same and denote this number by n. To test for an association between
the trait and the marker, we use the classical two-sample t-test. Let t denote the realization of the classical t-test statistic obtained from the formula (8.4), where the X_i,
$i \in \{1, \ldots, n\}$, denote the values of the trait for the individuals who are homozygous
at the marker and the Y_i, $i \in \{1, \ldots, n\}$, are the values of the trait for the individuals
who are heterozygous.

If our data contain no outliers and the sample size n is large enough (say $n > 30$),
we can use the central limit theorem and approximate the distribution of the t statistic
by the normal $N(\xi, 1)$ distribution, with noncentrality parameter

$$\xi = \frac{\mu_2 - \mu_1}{\sigma} \sqrt{\frac{n}{2}}. \tag{4.1}$$

Let us denote the normal approximation for the critical value of the two-sided t-test by
$t_c = \Phi^{-1}(1 - \alpha/2)$, where $\Phi(\cdot)$ denotes the cumulative distribution function of the
standard normal distribution and α is the assumed significance level. Straightforward
calculations lead to the following formula for the power of such a two-sided test:

$$P(\xi) \approx 1 - \Phi(t_c - \xi) + \Phi(-t_c - \xi).$$

4.1.2.2   When the Marker is at Some Distance from a QTL

Now, assume that a genetic marker is at a distance of d Morgans from a QTL. According to the Haldane mapping function, Eq. (2.2), the probability of recombination

between the QTL and the marker, is equal to $r = (1 - exp(-2d))/2$. Subsequently, the distribution of the trait values among individuals who are homozygous at the marker is a mixture of two normal distributions, namely,

$$T|M = AA \sim (1 - r)N(\mu_2, \sigma^2) + rN(\mu_1, \sigma^2).$$

Similarly,

$$T|M = aA \sim (1 - r)N(\mu_1, \sigma^2) + rN(\mu_2, \sigma^2).$$

Now, consider the statistic $D = \bar{X} - \bar{Y}$, where $\bar{X}$ is the mean value of the trait among individuals who are homozygous at a marker and $\bar{Y}$ is the mean value of the trait among those who are heterozygous. The expected value and variance of D are given as follows:

$$E(D) = (\mu_2 - \mu_1)(1 - 2r) \text{ and } Var(D) = \sqrt{\frac{2}{n}} \left(\sigma^2 + r(1 - r)(\mu_2 - \mu_1)^2\right).$$

This, in turn, leads to the following approximation for the power of the t-test statistic:

$$P(\delta_1) \approx 1 - \Phi(t_c - \delta_1) + \Phi(-t_c - \delta_1),$$

where $\delta_1 = \frac{(\mu_2 - \mu_1)(1 - 2r)}{\sqrt{\sigma^2 + r(1-r)(\mu_2 - \mu_1)^2}} \sqrt{\frac{n}{2}}$.

Figure 4.1 illustrates the power to detect a QTL as a function of the distance between the QTL and the marker when the backcross design is adopted. The graph

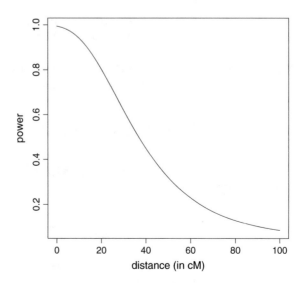

Fig. 4.1 The dependency of the power on the distance between the marker and QTL; $\mu_2 - \mu_1 = 1$, $\sigma = 1$, $n = 40$, $\alpha = 0.05$

illustrates the case where $\mu_2 - \mu_1 = 1$, $\sigma = 1$, and $n = 40$. When the marker is located at the QTL, the power of detection is equal to 0.994. As can be observed, the power systematically decreases as the distance between the marker and QTL increases. However, the QTL can be detected with power exceeding 80%, even when the marker is at a distance of 20 cM from the QTL.

4.1.3   Genome Wide Search with Tests Based on Single Markers

The rapid development of the field of QTL mapping started with the introduction of molecular markers, which enabled searching over the entire genome for genes associated with a given trait. The first methods for localizing a QTL relied on performing a sequence of tests for the association between a trait and a single marker. In typical genome scans using inbred populations, the number of markers reaches 500, which requires adjustment for multiple testing. Due to the strong correlation between the test statistics at neighboring markers, classical methods of multiple testing, such as the Bonferroni correction, are overly conservative. In [22, 26, 40], this problem was solved by observing that the trajectories of the t-test statistics evaluated at markers distributed over an entire chromosome can be viewed as realizations of the Ornstein–Uhlenbeck process. Based on such an approximation, the critical value t_c for the absolute value of the t-test statistic for detecting additive QTL effects using a backcross or intercross design can be numerically calculated from the following formula:

$$\alpha \approx 1 - \exp\left[-2C\left\{1 - \Phi(t_c)\right\} - 0.04Lt_c\phi(t_c)\nu\left(t_c\sqrt{0.04\Delta}\right)\right] = \alpha_{SM}(t_c), \quad (4.2)$$

where α is the target familywise error rate (FWER, the probability of detecting a QTL when in fact no QTL is present), C is the number of chromosomes, Δ is the distance between neighboring markers (in cM), L is the length of the chromosome (in cM), ϕ is the density function, Φ is the cumulative distribution function of the standard normal density, and $\nu(t)$ is given by the formula

$$\nu(t) = 2t^{-2}\exp\left\{-2\sum_{n=1}^{\infty} n^{-1}\Phi(-|t|n^{1/2}/2)\right\}. \quad (4.3)$$

According to Sect. 5.3 of [49], for numerical purposes it is sufficient to use the following approximation of $\nu(t)$:

$$\nu(t) \approx \frac{(2/t)(\Phi(t/2) - 0.5)}{(t/2)\Phi(t/2) + \phi(t/2)}. \quad (4.4)$$

Let $\mu = \mu_2 - \mu_1$ denote the effect of a QTL. The power to detect such an effect can be estimated according to the following formula (see Sect. 6.2 of [49]):

$$P(|T| > t_c) \approx 1 - \Phi(t_c - \xi) + \phi(t_c - \xi)\left[\frac{2\nu}{\xi} - \frac{\nu^2}{(t_c + \xi)^2}\right], \qquad (4.5)$$

where the noncentrality parameter ξ is given in (4.1) and

$$\nu = \nu\left(\sqrt{0.04t_c\Delta}\right).$$

Remark 4.1 Our noncentrality parameter slightly differs from the noncentrality parameter $\xi_1 = \sqrt{\frac{\pi}{2}}\frac{\mu}{\sqrt{\mu^2/4+\sigma^2}}$ from [49]. This results from using a different standardization of $D = \bar{X} - \bar{Y}$. To obtain the classical two-sample t-test statistic, used here, the standardization is performed with respect to the unbiased estimate of the variance of the error term σ^2. In [49], the difference between the means is standardized with respect to the total variance of the response variable. Under the null hypothesis of no QTL, these two approaches give very similar results. However, under the alternative, the realization of the classical t-test statistic is typically larger than the statistic proposed in [49] and thus has a larger power to detect a QTL.

Figure 4.2 compares the threshold calculated according to formula (4.2), the one based on the Bonferroni correction and the one obtained when no correction for multiple testing is used, together with the corresponding familywise error rates. It can be observed that ignoring any correction for multiple testing leads to a familywise error rate of almost 100% when the search is done using markers spaced every 1 cM and over 80% when one checks only five markers on each chromosome. This means that in at least 80% of experiments some false discoveries would be reported. The Bonferroni correction is too conservative for very densely spaced markers, leading to a familywise error rate which is substantially less than 0.05 when $\Delta \leq 5$ cM. When

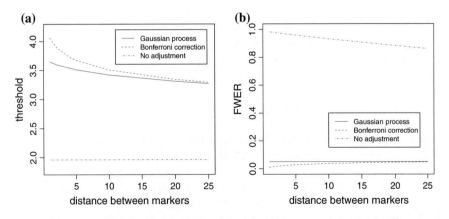

Fig. 4.2 The dependency of t-test thresholds and the familywise error rate as a function of the distance between markers: based on 10 chromosomes, each of length 100 cM; nominal FWER $\alpha = 0.05$

Fig. 4.3 Power curves;
$\mu = \mu_2 - \mu_1, \quad \sigma = 1,$
$n = 40, \quad \alpha = 0.05,$
$\Delta = 1\,\mathrm{cM}$

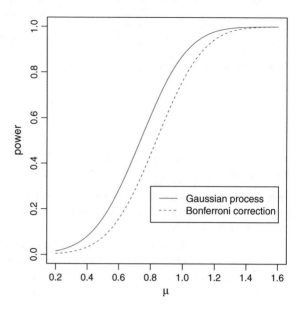

the markers are densely spaced, the thresholds given by (4.2) are substantially lower than the Bonferroni thresholds. As a result, for $\Delta = 1\,\mathrm{cM}$ the power to detect genes with moderately strong effects is on average 15 % larger than the power of the procedure based on the Bonferroni correction (see Fig. 4.3). The difference between the thresholds and powers for these two approaches diminishes as the distance between markers increases and is almost nonexistent for $\Delta = 25\,\mathrm{cM}$.

Due to the large correlation between neighboring markers, it often occurs that the QTL is not inferred to be at the closest marker. In [22], three different methods for constructing confidence regions for the location of a QTL were discussed and compared by means of simulation: a method based on the likelihood of change points, Bayes credible sets, and support regions, introduced in [17]. Based on this analysis, Dupuis and Siegmund [22] recommend support regions as being most robust with respect to changes in marker density and sample size. Using such an approach, the marker q falls within the support region of a QTL if

$$2 \ln LR(q) \geq \max_d 2 \ln LR(d) - x,$$

where x is a positive number, which determines the confidence level and $2 \ln LR(q)$ is the likelihood ratio statistic for detecting the presence of a QTL at position q. In the case of the backcross design, $2 \ln LR(q)$ roughly corresponds to the square of the t-test statistic. In the case of dense marker maps, in [22] it is reported that using $x = 6.6$ leads to support regions with a coverage probability of roughly 95 %. In the genetics literature, such intervals are called 1.5 LOD support regions, since the logarithm base 10 of the likelihood of the model falls by 1.5 units over such an interval.

Dupuis and Siegmund [22] also provide formulas for the lengths of the confidence/support regions. Their theory supports earlier empirical findings that the expected length of a confidence interval is roughly proportional to ξ^{-2}, where ξ is the noncentrality parameter, defined in (4.1). This implies that QTLs associated with larger effects can be located with greater precision and that the length of a support interval is roughly proportional to n^{-1}. This second result is interesting, since in most problems of classical statistics, the length of a confidence interval is proportional to $n^{-1/2}$. According to the discussion presented in [22], to obtain a precise estimate of the location of a QTL, one should use designs characterized by a relatively large recombination rate, such as recombinant inbred lines or advanced intercross. However, even in the case of such designs, the confidence interval for the location of a QTL usually spans several centimorgans. In the case of the backcross design, the confidence intervals for QTLs with a moderately strong effect can even span of 30–40 cM.

Permutation tests, discussed in Sect. 3.2.2, are another popular approach to correcting for the effects of multiple testing in the context of QTL mapping. Such tests were first proposed in [15]. Permutation tests do not require the assumption that the test statistics are normally distributed. However, one drawback is their large computational complexity. Moreover, due to the large sample sizes often used in QTL mapping, the distribution of the test statistics can usually be well approximated by the normal distribution. Thus, the theoretical results derived in [26] and [22] can often be used, even when a trait does not have a normal distribution.

4.2 Interval Mapping

4.2.1 Interval Mapping based on the mixture model

Interval mapping [40] allows us to estimate the location of a QTL positioned between genetic markers. To explain the idea of interval mapping, let us concentrate on the backcross design. In the case when the QTL locus coincides with the genetic marker, we can experimentally establish its genotypes X_i, $1 \leq i \leq n$ for all n individuals in the sample. If the QTL locus lies between two markers, we can use the information contained in the genotypes of the neighboring markers M_i to calculate the conditional probabilities $p_i = P(X_i = AA|M_i = m_i)$, where $m_i \in \{(AA, AA), (AA, aA), (aA, AA), (aA, aA)\}$ denotes the pair of genotypes of the ith individual at the neighboring markers. In this situation, we can often model the distribution of the trait using a mixture of normal distributions with different expected values:

$$f_i(y, \mu_1, \mu_2, \sigma^2) = p_i \frac{1}{\sqrt{2\pi}\sigma} \exp\left(-\frac{(y - \mu_2)^2}{2\sigma^2}\right)$$
$$+ (1 - p_i) \frac{1}{\sqrt{2\pi}\sigma} \exp\left(-\frac{(y - \mu_1)^2}{2\sigma^2}\right). \qquad (4.6)$$

The parameters (μ_1, μ_2, σ^2) can be estimated by maximizing the likelihood function

$$L_s(\mu_1, \mu_2, \sigma^2) = \prod_{i=1}^{n} f_i(Y_i, \mu_1, \mu_2, \sigma^2), \qquad (4.7)$$

where $Y_1, \ldots, Y_n$ are the trait values for each individual.

Since there are no closed-form analytical formulas for the values of the parameter maximizing L_s, estimators $(\hat{\mu}_1, \hat{\mu}_2, \hat{\sigma}^2)$ are calculated using a numerical method, in particular, the expectation maximization algorithm (EM, [21]).

The location of the QTL is estimated by maximizing the likelihood function over the set of possible QTL positions :

$$\hat{\delta} = \text{argmax}_{s \in \Lambda} L_s(\hat{\mu}_1, \hat{\mu}_2, \hat{\sigma}^2), \qquad (4.8)$$

where Λ is the set of locations on the chromosome which interest us.

Because we are usually unsure whether the considered part of the chromosome indeed contains a QTL, the null hypothesis $H_0 : \mu_1 = \mu_2$ needs to be tested using the likelihood ratio test

$$LR_\delta = 2 \log \left(\frac{L_\delta(\hat{\mu}_1, \hat{\mu}_2, \hat{\sigma}^2)}{L_\delta(\tilde{\mu}, \tilde{\sigma}^2)} \right),$$

where $\tilde{\mu} = \tilde{\mu}_1 = \tilde{\mu}_2$ and $\tilde{\sigma}^2$ denote the estimates calculated under the assumption that $\mu_1 = \mu_2$. To select the critical value c, so that the probability of a type I error (FWER) is controlled at the level α, the following formula from [44] can be used:

$$\alpha \approx 2\Phi(-\sqrt{c}) + \frac{2}{\pi} \exp(-c/2) \sum_{i=1}^{k} \arctan\left(\sqrt{r_i/(1 - r_i)}\right), \qquad (4.9)$$

where k is the number of intermarker intervals, and r_i is the probability of recombination on the ith interval.

However, in [48] it was observed that the above formula is only accurate when the distance between markers is relatively large (say 20 cM) and becomes conservative (overestimates α) in the case when the markers are more densely spaced. To solve this problem, Siegmund and Yakir [48] (see also [49]) proposed a new approximation, which in the case of equidistant markers combines Eqs. (4.2) and (4.9) into the formula

$$\alpha \approx \alpha_{SM}(\sqrt{c}) - 2\frac{L}{\Delta}\left[\frac{2\phi(\sqrt{c})}{\sqrt{2\pi}} \arctan\left(\sqrt{r/(1-r)}\right) - Pr(LR_0 \leq \sqrt{c}, LR_\Delta \geq \sqrt{c})\right],$$

(4.10)

where the function α_{SM} is defined in Eq. (4.2), L is the chromosome length, Δ is the distance between markers, r is the corresponding recombination fraction, and LR_0 and LR_Δ denote the values of the likelihood ratio statistics at the beginning of the chromosome and Δ cM from the beginning, respectively. According to the simulations in [49], this approximation performs very well and is much more accurate than Eq. (4.9) when the distance between markers is small. Alternatively, the permutation method described in [15] can be used to calculate the threshold c.

4.2.2 Regression Interval Mapping

In [31], a simpler version of interval mapping was proposed. This version is based on a simple regression model for QTL mapping.

To define a regression model for QTL mapping, we first need to define dummy variables which code the marker genotypes. In the case of the backcross design, we need just one such variable for a given marker, which can be defined as follows:

$$X_{ij} = \begin{cases} -1/2 \text{ if } Q_{ij} = AA \\ 1/2 \ \ \text{ if } Q_{ij} = Aa \end{cases},$$

(4.11)

where Q_{ij} corresponds to the genotype of the ith individual at the jth marker.

Now, a simple regression model to locate a QTL based on the backcross design can be written as follows:

$$Y_i = \beta_0 + \beta_j X_{ij} + \varepsilon_i,$$

(4.12)

where $Y_1, \ldots, Y_n$ are the trait values for n individuals, β_0 is the overall mean value of the trait, and $\beta_j = \mu_2 - \mu_1$ corresponds to the difference between the mean values of the trait according to the genotype at the jth marker. The parameters of the model (4.12) can be estimated using the least squares method and the null hypothesis of no QTL effect corresponds to the classical hypothesis of simple regression that $\beta_j = 0$ and can be tested using the t-test statistic for simple regression. As one would expect, this approach gives exactly the same results as the analysis based on the two-sample t-test described above. One major advantage of this approach is that the model (4.12) can be easily extended to include more markers and their interactions, which will be discussed in detail in the sections to follow.

When one wants to test for the presence of a QTL between markers, Haley and Knott [31] propose a very simple solution: they replace X_{ij} by its conditional expected value, based on the information included in the genotypes of neighboring markers. This version of interval mapping is much simpler than the method proposed in [40], since it does not require application of the EM algorithm. Results from simulations, e.g., in [37], show that, in many cases, the analysis of data using regression interval mapping gives almost exactly the same results as analysis based on the full mixture model (4.6). It is also expected that regression interval mapping is more robust with respect to the violation of assumptions regarding the distribution of the trait of interest.

Regression interval mapping can also be extended to the intercross design. In this case, we introduce an additional dummy variable to model the dominance effect:

$$Z_{ij} = \begin{cases} -1/2 & \text{if } Q_{ij} = AA \text{ or } aa \\ 1/2 & \text{if } Q_{ij} = Aa \end{cases}. \tag{4.13}$$

Now, the regression model can be extended to

$$Y_i = \beta_0 + \beta_j X_{ij} + \gamma_j Z_{ij} + \varepsilon_i. \tag{4.14}$$

Based on model (4.14), the hypothesis of no QTL influence takes the form $\beta_j = \gamma_j = 0$ and can be tested using the classical F test. In the case when one wants to test for the presence of a QTL between markers, the dummy variables X_{ij} and Z_{ij} can be replaced by their expected values based on the genotypes of the neighboring markers.

4.2.3 Nonparametric Version of Interval Mapping

In cases where the distribution of a trait differs substantially from the normal distribution, one can replace the t-test for the presence of a QTL by the Wilcoxon rank-sum test (see Sect. 8.7). In [39], this idea was extended to interval mapping and the following statistic for testing whether a QTL is present at the location s was proposed:

$$Y_w(s) = \sum_{i=1}^{n} [n + 1 - 2R_i] E[X_i(s)|DATA],$$

where R_i denotes the rank of the trait value for the ith individual among all trait values.

Let $Z_w(s) = \frac{Y_w(s)}{std(Y_w(s))}$ denote the standardized version of $Y_w(s)$. As discussed in [39], under the null hypothesis of no QTL, the distribution of $Z_w(s)$ converges to the standard normal distribution as the sample size n increases. Also, suppose s varies over the entire chromosome, the process $Z_w(s)$ converges to the Ornstein–Uhlenbeck process and correcting for the effects of multiple testing can be performed in the same way as discussed in Sects. 4.1.3 and 4.2.

In the case of the intercross design, this idea can be further extended using a rank regression test statistic (see Sect. 8.7.2) based on model (4.14). It can be shown that this approach is asymptotically equivalent to performing a classical F test, where the trait values Y_i are replaced by their ranks R_i.

4.2.4 Specific Models

Nonparametric tests, mentioned in the previous paragraph, enable the statistical local-ization of QTLs in the case when the distribution of a trait is not normal. However, they are really distribution free only when a trait is continuous (or, in other words, the probability of observing the same value more than once is equal to zero). Instead, in the case when the trait of interest is categorical, we can use one of the many models designed for the analysis of discrete data. Specifically, for binary traits a search over single markers can be performed using classical chi-square tests for independence (see Sect. 8.8.2) or likelihood ratio tests based on logistic regression (see Sect. 8.5), where the predictors consist of marker genotypes, coded as in (4.11) and (4.13). Similarly, in the case of count data one can carry out Poisson regression or one of its generalizations: Generalized Poisson regression, which allows some flexibility in estimating the dispersion, or zero-inflated generalized Poisson regres-sion (ZIGPR), which assumes that there exists some additional probability mass at zero (see Sect. 8.5.1 for a discussion on generalized linear models). Similar to the approach used to define ZIGPR, in the context of QTL mapping researchers sometimes model the distribution of a quantitative trait using a mixture of a con-tinuous distribution (e.g., normal) and an additional "spike" at zero (see e.g., [14]). Such a model is appropriate for modeling, e.g., tumor size, since healthy individuals simply do not have a tumor. All of these probabilistic models specify the relation between the density/probability mass of the trait and the marker genotype and can be easily extended to interval mapping. We refer the reader to [55] for a description of the application of interval mapping to binary traits based on logistic regression or to [19, 20] for a description of interval mapping based on generalized Poisson regression and zero-inflated generalized Poisson regression, respectively. In many applications of these models to QTL mapping, the critical values required for inferring the presence of a QTL are calculated using the permutation method. It is, however, important to note that in most cases the asymptotic distribution of the corresponding likelihood ratio statistic is chi-square and therefore the corresponding critical values can often be well approximated by the appropriate values for the likelihood ratio tests under the idealized normal model, as discussed in Sects. 4.1.3 and 4.2.

4.2.5 Overestimation of Genetic Effects

If the location of a QTL is known, then the estimators of the parameters (μ, β, σ) in the regression model (4.12) are approximately unbiased. Specifically, one can accurately estimate the QTL's effect size β. This is no longer true when a researcher performs QTL mapping to infer the presence of a QTL and estimate its location δ. In this case, δ is estimated according to formula (4.8) and the hypothesis $H_0 : \beta = 0$ is tested by comparing $L_{\hat{\delta}}$ with the corresponding critical value. If H_0 is rejected, then

$\hat{\delta}$ is the estimator of the QTL's location and the estimate of β in the corresponding simple regression model is used as the estimate of the QTL's effect.

It is well known that when one estimates β only in the case when H_0 is rejected, one tends to overestimate the genetic effect (see e.g., [1, 13, 29, 56]). Furthermore, another source of bias is identified in [9]. Due to the random nature of data, $\hat{\delta}$ often differs from the true location of a QTL. Since $\hat{\delta}$ is defined to be the point at which the estimated effect of a QTL attains its maximal value, the corresponding value of $\hat{\beta}$ is typically larger than the estimator of β based on the actual location of the QTL. The extensive simulation study reported in [9] illustrates that inferring the existence of a QTL and estimating its location have comparable influences on the overestimation of a QTL's effect. This bias is increasing in the length of the portion of the genome searched and decreasing in both the sample size and the true effect size. This is due to the fact that for large sample sizes and large gene effects it is easier to reject H_0 (so testing no longer has a large influence on the outcome of the estimation procedure) and the estimator of a QTL's location is very precise.

4.3 Model Selection

4.3.1 QTL mapping with mBIC

The methods of gene mapping discussed above are based on a simple statistical model, which assumes that the trait of interest is influenced by just one QTL. In reality, however, many quantitative traits are influenced by several major QTLs, which may interact with each other. In the case of a continuous trait whose distribution can be approximated by the normal distribution, such a situation can be well described by a multiple regression model. Due to the large computational complexity involved in analyzing genome scans, researchers usually only search for interactions of order 2. In this case, one can define a multiple regression model for QTL mapping based on backcross or recombinant inbred lines in the following form:

$$Y_i = \beta_0 + \sum_{j=1}^{m} \beta_j X_{ij} + \sum_{1 \leq j < l \leq m} \gamma_{jl} X_{ij} X_{il} + \varepsilon_i, \tag{4.15}$$

where $X_{ij} \in \{-1/2, 1/2\}$ represents the genotype of the ith individual at the jth QTL and $\varepsilon_i \sim N(0, \sigma^2)$ models the random noise.

Similarly, in the case of the intercross design, one can use Cockerham's coding (see Table 2.3) to build a multiple regression model of the form

$$Y_i = \beta_0 + \sum_{j=1}^{m} \left(\beta_j X_{ij} + \delta_j Z_{ij} \right)$$

$$+ \sum_{1 \le j < l \le m} \left(\gamma_{x_j x_l} X_{ij} X_{il} + \gamma_{x_j z_l} X_{ij} Z_{il} + \gamma_{z_j x_l} Z_{ij} X_{il} + \gamma_{z_j z_l} Z_{ij} Z_{il} \right) + \varepsilon_i. \quad (4.16)$$

In the above formulas, the first sum describes the main effects of single QTLs, while the second sum describes the interactions between pairs of genes. In the case of the backcross and intercross designs, the proposed coding of genotypes avoids additive and interaction effects becoming confounded. It is also important to note that some of the regression coefficients can be equal to zero, which enables modeling situations where some genes do not interact with others, or some interacting genes do not have their own main effects. Below, we will use k_1 to denote the number of nonzero main effects (additive and dominant) and k_2 to denote the number of nonzero interaction terms.

To explain the basic idea of multiple QTL mapping, we start by deriving a model in which each QTL is represented by its closest marker. Thus our task reduces to identifying the best model of the form

$$Y_i = \mu + \sum_{j \in I} \beta_j X_{ij} + \sum_{(u,v) \in U} \gamma_{uv} X_{iu} X_{iv} + \varepsilon_i, \quad (4.17)$$

where X_{ij} denotes the genotype of the ith individual at the jth marker, I is a certain subset of $\mathcal{N} = \{1, \ldots, N_m\}$, N_m is the number of available markers, and U is a certain subset of $\mathcal{N} \times \mathcal{N}$. This can be solved by applying criteria for model selection, as discussed in Chap. 3. This problem is nontrivial, since in the majority of applications the number of markers, N_m, is larger or comparable to the sample size n. Moreover, the complexity of the problem is hugely increased by considering interaction effects, whose number is of order N_m^2. This places QTL mapping in the context of a high-dimensional problem of model selection, where classical tools, such as the Bayesian information criterion (3.28), have a strong tendency to choose overly complicated models (e.g., see [12]). To solve this problem in [6], a specific version of the modified BIC (see (3.29)) was proposed. When deriving this criterion, it is assumed that all the main effects have the same prior probability, ν_1, of appearing in the true multiple regression model independently of other components. Similarly, let ν_2 be the probability that any given interaction term appears in the true model. Let M be a model containing k_1 main components and k_2 interactions. Our assumptions imply that the prior probability of M equals

$$\pi(M) = \nu_1^{k_1} \nu_2^{k_2} (1 - \nu_1)^{N_m - k_1} (1 - \nu_2)^{N_e - k_2},$$

where $N_e = \binom{N_m}{2}$. This implies that the prior distributions of the number of main and interaction effects are binomial with expected values $E_m = N_m \nu_1$ and $E_e = N_e \nu_2$, respectively.

After adapting the BIC on the basis of this prior probability and eliminating the terms which do not depend on the selected model, we obtain the modified Bayesian information criterion

$$mBIC = BIC - 2\log \pi(M)$$

$$= n\log RSS + (k_1 + k_2)\log n + 2k_1 \log \left(\frac{N_m}{E_m} - 1\right) + 2k_2 \log \left(\frac{N_e}{E_e} - 1\right), \quad (4.18)$$

where RSS is the residual sum of squares for the model under consideration.

In the case where there are no prior expectations concerning the number of QTLs, Bogdan et al. [6] proposed calibrating this criterion, so that it controls the familywise error rate in the weak sense (the probability of detecting at least one QTL when in fact no gene influences the response) at a level below 10 % when the sample size $n \geq 200$. As a result of the extensive simulations and theoretical calculations presented in [6], the default values $E_m = E_e = 2.2$ were proposed. Hence, the basic version of mBIC for identifying main and two-way interaction effects based on the backcross design minimizes the following objective function:

$$mBIC_S = n\log RSS + (k_1 + k_2)\log n + 2k_1 \log \left(\frac{N_m}{2.2} - 1\right) + 2k_2 \log \left(\frac{N_e}{2.2} - 1\right). \quad (4.19)$$

The desirable properties of $mBIC_S$ have been illustrated by an extensive simulation study reported in [6], which shows that this criterion offers a very good compromise between the power to detect influential genes and control of the number of false discoveries. Furthermore, in [3] an extension of mBIC that adapted the intercross design was proposed. Since in the multiple regression model corresponding to the intercross design, (4.16), one needs to consider both additive and dominance terms, the number of possible main effects is equal to $2N_m$, and the number of possible two-way interactions is $4\binom{N_m}{2} = 2N_m(N_m - 1)$. Hence, the corresponding version of mBIC proposed in [3] takes the form

$$mBIC = n\log RSS$$
$$+ (k_1 + k_2)\log n + 2k_1 \log(2N_m/2.2 - 1) + 2k_2 \log(2N_m(N_m - 1)/2.2 - 1). \quad (4.20)$$

Additionally, in [3] a new, two-stage version of mBIC was proposed. This criterion adapts the basic form of mBIC (4.18), which in the case of the intercross design takes the form

$$mBIC_1 = n\log RSS + (k_1 + k_2)\log n + 2k_1 \log \left(\frac{2N_m}{E_m} - 1\right) + 2k_2 \log \left(\frac{2N_m(N_m - 1)}{E_e} - 1\right), \quad (4.21)$$

where, as in (4.18), E_m and E_e denote the expected numbers of real main and interaction terms. In the first step of the two-stage procedure, the numbers of main effects and interactions are estimated by the respective numbers of regression terms in the model selected according to the standard version of mBIC (4.20). Then the search procedure is repeated using a criterion of the form (4.21), where the expected values

E_m and E_e are given by the estimates obtained in the first step. The theoretical calculations and simulations reported in [3] demonstrate that this two-stage version of mBIC controls the false discovery rate at a reasonably low level, while providing a slightly larger power than the standard version of mBIC.

4.3.2 Robust Version of mBIC

In [3, 6], mBIC was derived under the assumption that the random terms in these multiple regression models have a normal distribution. This assumption is, however, not crucial. It is now well known that standard least squares analysis of regression models is robust to moderate deviations from normality, as long as the sample size n is large enough. This observation is justified by an extension of the central limit theorem (see Sect. 6.1), which says that under a variety of error distributions, the least squares estimators of the coefficients in the multiple regression model have an asymptotically normal distribution. However, there exist deviations from normality where the central limit theorem can no longer be relied upon. For example, if the distribution of the error term has very heavy tails, this theorem might not even hold. Such a situation occurs when the error terms have a Cauchy distribution, as mentioned in Sect. 6.2.2. Standard regression used in conjunction with mBIC is also unreliable in the case where the data contain a certain proportion of outliers, which have a strong influence on the residual sum of squares. This problem was addressed in [4], where several robust versions of mBIC were proposed and carefully investigated.

In the rich literature on robust regression (see e.g., [35]), the residual sum of squares $RSS = \sum_{i=1}^{n} r_i^2$ is usually replaced by $\sum_{i=1}^{n} \rho(\tilde{r}_i)$, where the $\tilde{r}_i$ are residuals which are standardized using a robust scale estimator and $\rho(x)$ is a contrast function. These contrast functions are selected in such a way that they reduce the influence of outliers as compared to the analysis based on the standard squared error function. In [4], three of the most popular contrast functions are applied:

$$
\rho_{\text{Huber}}(x) = \begin{cases} c|x| - c^2/2 & \text{for } |x| > c \\ x^2/2 & \text{for } |x| \leq c, \end{cases}
$$

$$
\rho_{\text{Hampel}}(x) = \begin{cases} a(b - a + c)/2 & \text{for } \quad |x| > c \\ a(b - a + c)/2 - \frac{a(|x|-c)^2}{2(c-b)} & \text{for } b < |x| \leq c \\ a|x|^2 - a^2/2 & \text{for } a < |x| \leq b \\ x^2/2 & \text{for } \quad |x| \leq a. \end{cases}
$$

$$
\rho_{\text{Bisquare}}(x) = \begin{cases} c^2/6 & \text{for } |x| > c \\ c^2/6\left[1 - \left(1 - \left(\frac{x}{c}\right)^2\right)^3\right] & \text{for } |x| \leq c. \end{cases}
$$

The Hampel and Bisquare contrast functions are bounded, and thus they substantially reduce the influence of outliers on the outcome of statistical analysis. For large x, the Huber contrast function is linear, which reduces the influence of outliers as compared to the standard squared error function. When c converges to 0, robust analysis based on the Huber contrast function converges to so-called L_1 regression, where the parameters are chosen to minimize $\sum_{i=1}^{n} |r_i|$. This is another very popular method for the robust analysis of regression data.

The calculation of regression coefficients based on these contrast functions requires iterative method, such as iteratively reweighted least squares. In [4], the residuals were standardized using the median absolute deviation (MAD) from the median. For details on this and other aspects of robust regression (e.g., confidence regions and tests of significance), we refer the reader to Chap. 7 in [34].

A natural way to obtain a robust version of BIC (or mBIC) is to replace the residual sum of squares in the criterion for model selection by a sum of contrasts. However, unlike least squares regression, a sum of contrasts will usually not be scale invariant. Therefore, in [4] the values of the response variable Y_i are standardized in a robust way by subtracting the median and dividing by the MAD. Let $Y_i^{(s)}$ denote the values of the trait after this rescaling procedure. For the purpose of estimating the parameters of various regression models, their residuals are additionally rescaled (separately for each model) by the robust rescaling provided by the function rlm in the MASS library of the R package, which is available from http://www.R-project.org.

Our goal is to construct a robust version of mBIC, so that it controls FWER at the desired level. This requires rescaling the corresponding robust version of the likelihood ratio statistic so that it has an asymptotic chi-square distribution. To understand this procedure, let us consider the following two multiple regression models: M_2, containing $k_{1,2}$ main effects and $k_{2,2}$ interactions, and a nested model M_1, containing $k_{1,1}$ main effects and $k_{2,1}$ interactions. To compare these models, we define the following robust version of the likelihood ratio statistic:

$$D_n = n(\log \sum \rho(Y_i^{(s)} - x_i'\hat{\theta}_1) - \log \sum \rho(Y_i^{(s)} - x_i'\hat{\theta}_2)),$$

where $\tilde{\theta}_1 = (\theta_{1,1}, \ldots, \theta_{1,k_{1,1}+k_{2,1}})$ and $\tilde{\theta}_2 = (\theta_{2,1}, \ldots, \theta_{2,k_{1,2}+k_{2,2}})$ are the vectors of parameters used in M_1 and M_2, and $\hat{\theta}_1$ and $\hat{\theta}_2$ are their robust estimators.

The following theorem is proved in [4].

Theorem 4.1 *Let $\tilde{Y}_i^{(s)}$ be the observations standardized according to the population median and the population MAD. Moreover, let $\rho(\cdot)$ be a contrast function satisfying the regularity conditions specified in Sect. 5.5 of [35] and let $\psi(\cdot)$ be the corresponding score function $\psi(x) = \rho'(x)$. Additionally, we define $\gamma = \int \psi'(x)f(x)dx$, $\sigma_\psi^2 = \int \psi(x)^2 f(x)dx$ and $\delta = \int \rho(x)f(x)dx$, where $f(x)$ denotes the density of the distribution of the random errors. Let $c_e = 2\gamma\delta/\sigma_\psi^2$.*
If the model M_1 holds, then

$$c_e D_n \xrightarrow{d} \chi^2_{(k_{1,2}+k_{2,2})-(k_{1,1}+k_{2,1})}, \tag{4.22}$$

when $n \to \infty$.

The constant c_e depends on the distribution of the error terms. Since in practical applications this distribution is often unknown, in [4] the following estimation procedure is proposed. In the first step, the "best" regression model is selected using the appropriate version of mBIC (see 4.23) with c_e set to 1, i.e., the normalizing constant for Gaussian errors. The empirical distribution of the resulting residuals is then used to approximate the expected values, estimating $\psi'(x)$, $\psi(x)$ and $\rho(x)$ by the corresponding averages. The estimator of c_e calculated this way will be denoted by $\hat{c}_e$.

The final robust version of mBIC suggests selecting the model which minimizes

$$mBIC = \hat{c}_e n \log \sum \rho(Y_i^{(s)} - x_i'\hat{\theta}) + (k_1 + k_2) \log n$$
$$+ 2k_1 \log \left(\frac{N_m}{2.2} - 1\right) + 2k_2 \log \left(\frac{N_e}{2.2} - 1\right). \qquad (4.23)$$

The extensive simulation study reported in [4] illustrates the desirable properties of the robust versions of mBIC. These robust criteria yield results comparable to those of standard mBIC when the error terms have a normal distribution and provide a much larger power when the data contain a certain proportion of outliers. However, the large computational complexity of such robust procedures for QTL mapping is a drawback.

4.3.3 Version of mBIC Based on Rank Regression

In [58], another robust version of mBIC, based on rank regression, is proposed. Under this version of mBIC, rBIC, the original values of the response variable are replaced by their corresponding ranks. Selection of the optimal model is then performed according to the standard version of mBIC. In the case when the error term comes from a continuous distribution, then under the null hypothesis of no QTL the distribution of rBIC does not depend on the distribution of the error term. Thus it is possible to rescale the criterion so that it controls FWER in the weak sense, independently of the unknown distribution of errors. Moreover, in [58] it is proved that when the sample size increases, the corresponding version of the likelihood ratio statistic has an asymptotic chi-square distribution. Therefore, the penalty which controls FWER for mBIC can be directly used for rBIC.

Let $R_i \in \{1, \ldots, n\}$ denote the rank of Y_i in the vector $Y = (Y_1, \ldots, Y_n)'$ and $R = (R_1, \ldots, R_n)'$ be the corresponding vector of ranks. Now, let us define the rank residual sum of squares as $rRSS = \sum_{i=1}^{n}(R_i - \tilde{R}_i)$, where $\tilde{R}_i = X(X'X)^{-1}X'R$. The rank version of mBIC takes the form

$$rBIC_S = n \log rRSS + (k_1 + k_2) \log n + 2k_1 \log \left(\frac{N_m}{2.2} - 1\right) + 2k_2 \log \left(\frac{N_m(N_m - 1)}{4.4} - 1\right)$$

for the backcross design and

$$rBIC = n \log rRSS$$
$$+ (k_1 + k_2) \log n + 2k_1 \log \left(\frac{2N_m}{2.2} - 1 \right) + 2k_2 \log \left(\frac{2N_m(N_m - 1)}{2.2} - 1 \right)$$

for the intercross design. The extensive simulations and analysis of real data reported in [58] demonstrate that rBIC indeed controls FDR at the assumed level and yields a high power, independently of the distribution of the error term. According to this simulation study, rBIC performs similarly to the robust versions of mBIC based on the contrast functions discussed above. The great advantage of rBIC is its ease of implementation and low computational cost.

4.3.4 Extensions to Generalized Linear Models

Modified versions of BIC can be also applied when the trait Y has a discrete distribution. In this situation, the dependence between the phenotype and marker genotypes is usually described by a generalized linear model (gLM), discussed in Sect. 8.5. One classical example is logistic regression, often used to model binary traits, such as disease status. Another popular model is Poisson regression, often applied to modeling count data, such as number of drinks per day or number of gallstones. In this case, mBIC can be written in its general form (see Eq. 3.29), with $n \log RSS$ replaced by $-2 \log \mathcal{L}(\hat{\theta}|Y)$, where $\mathcal{L}(\theta|Y)$ denotes the likelihood of the data under a given GLM and $\hat{\theta}$ is the maximum likelihood estimator of the vector of model parameters. The simulation study reported in [59] demonstrates that mBIC and mBIC2 possess

Table 4.1 Average number of false positives (FP), power and false discovery rate (FDR) using mBIC to select various classes of models and $n = 200, 500$ and $\omega = 20\%, 40\%$

	$n = 200$									
	$\omega = 20\%$					$\omega = 40\%$				
	LS	PoiR	ZIPR	GPR	ZIGPR	LS	PoiR	ZIPR	GPR	ZIGPR
TP	3.181	8.172	7.823	2.706	4.986	1.846	8.008	7.154	0.665	3.750
FP	0.125	21.075	11.309	0.658	0.357	0.106	27.033	9.614	0.296	0.373
Power	0.318	0.817	0.782	0.271	0.499	0.185	0.801	0.715	0.066	0.375
FDR	0.036	0.711	0.575	0.188	0.062	0.050	0.764	0.558	0.282	0.088
	$n = 500$									
	$\omega = 20\%$					$\omega = 40\%$				
	LS	PoiR	ZIPR	GPR	ZIGPR	LS	PoiR	ZIPR	GPR	ZIGPR
TP	5.282	9.143	9.095	4.540	6.674	4.168	8.970	8.725	2.024	6.187
FP	0.112	24.662	14.818	0.642	0.215	0.117	30.817	13.813	0.405	0.234
Power	0.528	0.914	0.909	0.454	0.667	0.417	0.897	0.873	0.202	0.619
FDR	0.019	0.723	0.607	0.112	0.028	0.025	0.770	0.599	0.142	0.033

desirable properties in problems involving the selection of appropriate logistic or Poisson regression models. Moreover, in [25] *mBIC* was applied to the problem of identifying genes influencing count traits using zero-inflated generalized Poisson regression (see Sect. 8.5.1). Table 4.1 provides some results of the simulation study from [25], which involved such QTL mapping based on a backcross mouse population. The mean number of false positives *FP*, power, and *FDR* is calculated based on 1000 simulations of the experiment, where the trait values were generated from the ZIGPR model, $Y_i \sim ZIGP(\mu_i(\beta), \varphi, \omega))$ (see Eq. 6.13) with

$$\mu_i(\beta) := \exp\left\{2.05 + X'_{Q,i}\beta\right\}, \quad i = 1, \ldots, n. \tag{4.24}$$

Here, $X_{Q,i} = (X_{Q1,i}, \ldots, X_{Q10,i})'$ denotes a vector of genotypes at ten approximately unlinked QTLs (see [25] for the exact location of the QTLs), with homozygotes and heterozygotes being coded as $-1/2$ and $1/2$, respectively, and the parameter values chosen to be

$$\beta = (-0.20, 1.00, 0.25, -0.60, 0.80, 1.20, 0.70, -0.15, -0.40, 1.50)'.$$

Additionally, $\varphi = 2$ and the zero-inflation parameter $\omega \in \{20\,\%, 40\,\%\}$. Table 4.1 compares the analysis based on the correct ZIGPR model with the analyses based on simpler models: Poisson regression (PoiR), generalized Poisson regression (GPR), and zero-inflated Poisson regression (ZIPR). Moreover, we report the results of the analysis of these simulated data using classical least square regression (LS) and mBIC to select a model.

The simulations reported in Table 4.1 show that using mBIC based on the correct ZIGPR model provides a low false discovery rate, while maintaining a relatively high power. Interestingly, the second best procedure is mBIC based on least squares regression, LS. While LS clearly performs worse than the correct ZIGPR model, it outperforms other misspecified models based on Poisson regression. Specifically, LS offers a much larger power than generalized Poisson regression (GPR) without a parameter for zero-inflation. Under the models which do not consider an overdispersion parameter, PoiR and ZIPR, we observe the opposite, i.e., they offer a much higher power than mBIC based on LS, or even ZIGPR, but this is at the cost of a very high false positive rate. The FDR of PoiR and ZIPR systematically exceeds 50 %, which implies that the number of false positives usually exceeds the number of true discoveries. This phenomenon is very interesting, since according to the theoretical results and simulation studies reported in [59], using mBIC to select a PoiR model performs very well with respect to FDR when the data are generated under the assumptions of Poisson regression. Also, as reported in [25] under the total null hypothesis, mBIC with PoiR controls FWER at the assumed level. We believe that when the data are generated under the assumptions of ZIGPR, the criteria based on PoiR and ZIPR pick too many regressors in order to account for the heterogeneity in the data caused by overdispersion.

4.3.5 mBIC for Dense Markers and Interval Mapping

In [8], it is observed that the penalty in mBIC based on the number of regressors is closely related to the Bonferroni correction for multiple testing. As already noted in Sect. 3.2.1, the Bonferroni correction keeps FWER close to the assumed level in the case where the test statistics for different tests are independent of each other. In the case of QTL mapping, such a situation only occurs when the markers are distant from each other, so their genotypes can be modeled as independent random variables. However, if markers are densely spaced, the state variables describing their genotypes are strongly correlated and the additional penalty applied in mBIC is excessive. This results in a low number of false discoveries, but also in an unnecessary loss of power to detect genes with significant effects. In [7], this problem was addressed and a new method for rescaling the penalty in mBIC was proposed.

To understand the method proposed in [7], let us consider the classical single marker regression model for the backcross design (4.12). Denote by LR_j the likelihood ratio statistic for testing the hypothesis $H_{0j} : \beta_j = 0$, i.e.,

$$LR_j = 2\log\left\{L(Y|\hat{\beta}_j)/L(Y|\beta_j = 0)\right\}.$$

Let us now consider a sequence of N_m markers located on a single chromosome, such that the distance between neighboring markers is equal to Δ cM. Moreover, let c denote the critical value for the likelihood ratio statistic such that FWER in the weak sense is controlled at the level α for the search over this chromosome, i.e.,

$$P_{H_0}\left(\max_{j\in\{1,\ldots,N_m\}} LR_j > c\right) = \alpha. \tag{4.25}$$

Here H_0 denotes the null hypothesis that none of the markers are associated with the trait.

Now, let us denote by N^{eff} the number of independent likelihood ratio statistics $\widetilde{LR}_1, \ldots, \widetilde{LR}_{N^{eff}}$ with exact chi-square distribution χ_1^2 such that the probability of their maximum exceeding c is approximately equal to α, i.e.,

$$\alpha = P_{H_0}\left(\max_{j\in\{1,\ldots,N^{eff}\}} \widetilde{LR}_j > c\right) = 1 - \prod_{i=j}^{N^{eff}}\left\{1 - Pr(\widetilde{LR}_j > c)\right\}$$

$$\approx 1 - \left[1 - 2\left\{1 - \Phi\left(\sqrt{c}\right)\right\}\right]^{N^{eff}}. \tag{4.26}$$

Now, comparing Eqs. (4.25) and (4.26), we can calculate the corresponding number of "effective" tests as

$$N^{eff} = \log(1-\alpha)/\log\{2\Phi(\sqrt{c}) - 1\}. \tag{4.27}$$

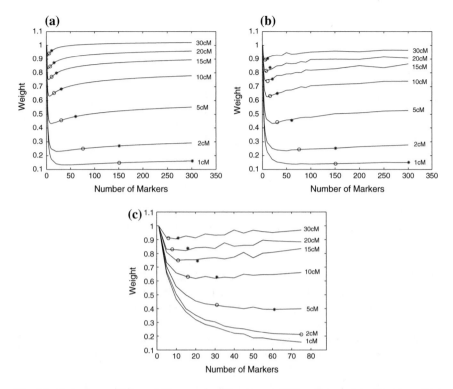

Fig. 4.4 Dependence of the weights corresponding to $\alpha = 0.05$ on the number of markers (N_m) and the genetic distance between markers (Δ). For additive effects, both theoretical and simulation results are reported. Circles and stars denote chromosomes of length 150 and 300 cM, respectively. w_{SM}^{add} theoretical (**a**), w_{SM}^{add} simulated (**b**), w_{SM}^{int} simulated (**c**)

Applying the mBIC in the case of dense markers, as in [7], the numbers of possible markers and one-way interactions, N_m and N_e, are replaced by the corresponding numbers of effective tests, N_m^{eff} and N_e^{eff}.

The basic difficulty of applying this methodology lies in the calculation of the critical value c. In the case of main effects, c can be calculated based on the theoretically derived formula (4.2), where $c = t_c^2$. In the case of interactions, c can be derived by computer simulations, as described in [7].

To aid application of the proposed criterion for model selection, Bogdan et al. [7] provide "weights" $w(\Delta, N_m)$, used to rescale the criterion when a search is based on N_m markers equidistantly placed at a distance of ΔcM

$$w(\Delta, N_m) = N^{eff}/N_m. \tag{4.28}$$

Figure 4.4 compares the weights for additive effects, w_{SM}^{add}, where the constant $c = t_c^2$ was obtained both from formula (4.2) and from computer simulations. It also presents simulated weights for interactions, w_{SM}^{int}. It can be observed that the weights

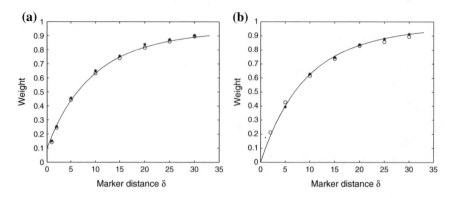

Fig. 4.5 Comparison of weights obtained by simulation with their estimating functions. The circles and stars denote chromosomes of length 150 and 300 cM, respectively. **a** $\hat{w}_{SM}^{add}$, **b** $\hat{w}_{SM}^{int}$

clearly depend not only on Δ, but also on the number of markers on a chromosome. It is evident that for any fixed Δ, $w_{SM}^{add}(\Delta, 1) = 1$. Note that $w_{SM}^{add}(\Delta, N_m)$ rapidly decreases (as a function of N_m), before finally starting to increase. However, Fig. 4.4a, b demonstrates that, for moderate and large values of N_m, the rate of increase of w_{SM}^{add} is very slow, and the weights remain stable over a wide range of N_m. Figure 4.4a, b also shows that the theoretical weights obtained from formula (4.2) are very close to the weights obtained by simulations.

Moreover, in [7] numerical approximation was used to derive estimates of the proposed weights as functions of the distance between markers, Δ:

$$\hat{w}_{SM}^{add}(\Delta) = 1 - 0.9e^{(-10\Delta/100+10(\Delta/100)^2)} \qquad (4.29)$$

and

$$\hat{w}_{SM}^{int}(\Delta) = 1 - e^{(-10.7\Delta/100+8.7(\Delta/100)^2)}. \qquad (4.30)$$

As shown in Fig. 4.5, these functions give very accurate approximations of the simulated weights. In the case where markers are not equally spaced, Bogdan et al. [7] suggest replacing Δ with the average distance between neighboring markers. The corresponding version of mBIC involves minimizing the following function:

$$mBIC_S = n \log RSS + (k_1 + k_2) \log n$$
$$+ 2k_1 \log \left(w_{SM}^{add} N_m/2.2 - 1\right) + 2k_2 \log \left(w_{SM}^{int} N_e/2.2 - 1\right).$$

The calculations presented in the Appendix of [7] demonstrate that the proposed version of mBIC controls FWER in the weak sense at the assumed level. The effectiveness of this version of mBIC has also been confirmed by extensive simulation studies and the analysis of real data.

In the case where the markers are widely spaced, methods for localizing QTLs between markers use interval mapping (see Sect. 4.2) or its multivariate extension, multiple interval mapping [36]. In [7], Bogdan et al. propose an adaptation of mBIC to widely spaced markers based on the multivariate version of regression interval mapping defined in [31].

Multiple regression interval mapping (MRIM) involves fitting the multiple regression model

$$Y_i = \mu + \sum_{j \in I} \beta_j G_{ij} + \sum_{u,v \in U} \gamma_{uv} G_{iu} G_{iv} + \varepsilon_i, \tag{4.31}$$

based on a dense set of possible QTL locations. Here, G_{ij} is the expected value of the variable coding the genotype of the ith individual at the jth position on the genome, conditional on the genotypes of the neighboring markers (the formulas for G_{ij} are given in [37]), and I and U are the sets of locations corresponding to QTLs with significant main effects and epistatic effects, respectively.

The main difference between MRIM and searching over individual markers is that a much larger set of possible QTL locations is investigated using MRIM. To take this increased number of possible locations into account, the penalty in the mBIC criterion needs to be increased. Because the predictor variables corresponding to the given intermarker locations are completely specified by the genotypes of the neighboring markers, the correlations between the likelihood ratio statistics at neighboring locations are stronger than when searching over a dense map of markers. To adapt mBIC to MRIM, the effective number of tests N_{IM}^{eff} corresponding to a genome search based on one-dimensional interval mapping needs to be calculated. To compute the critical value c for interval mapping, Bogdan et al. [7] use formula (4.9). Subsequently, the effective number of independent tests N_{IM}^{eff} is obtained using Eq. (4.27). The corresponding weight for additive effects is $w_{IM}^{add} = N_{IM}^{eff}/N$.

Figure 4.6a, b compares these weights with the weights simulated on the basis of regression interval mapping, where statistical tests for the presence of a QTL are performed at steps of 2 cM. Interestingly, these simulated weights are systematically smaller than the theoretical ones. Especially, large differences occur when the distance between neighboring markers is small. This agrees well with the discussion in [48, 49], which indicates the conservativeness of approximation (4.9), and calls for the application of formula (4.10) in future research. Figure 4.6 also presents simulated weights for the interaction terms. Here, we observe a relatively large difference in weights for chromosomes of length 150 and 300 cM, especially when the distance between markers exceeds 20 cM. However, it is worth mentioning that even these relatively large differences have a rather small influence on the application of mBIC, which depends on these weights through their logarithms.

For the purpose of applying mBIC$_{IM}$ to real data, the following empirical approximations of the weights for additive and interaction terms are proposed in [7]:

$$\widehat{w}_{IM}^{add}(\Delta) = -0.15 + 3.1\sqrt{\Delta/100} - 1.3\Delta/100 \tag{4.32}$$

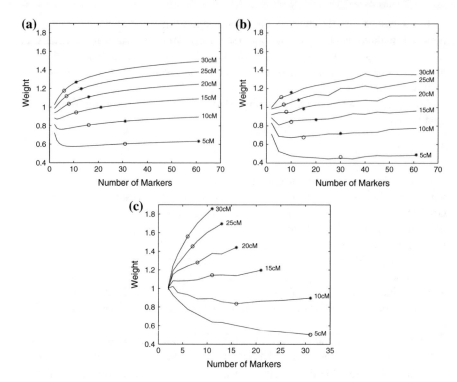

Fig. 4.6 Dependence of the interval marker weights corresponding to $\alpha = 0.05$ on the number of markers (N_m) and the genetic distance between markers (Δ). For additive effects, both theoretical and simulation results are reported. The circles and stars denote chromosomes of length 150 and 300 cM, respectively. **a** w_{SM}^{add} theoretical, **b** w_{SM}^{add} simulated, **c** w_{SM}^{int} simulated

and

$$\widehat{w}_{IM}^{epi}(\Delta) = -0.53 + 5.4\sqrt{\Delta/100} - 2.7\Delta/100, \qquad (4.33)$$

where the approximation for w_{IM}^{epi} is tailored to chromosomes of approximately 150 cM in length.

The accuracy of these approximations is illustrated in Fig. 4.7. In situations where markers are not equally spaced, the average distance between markers $\bar{\Delta}$ can be used to yield weights $w_{IM}^{add} = \widehat{w}_{IM}^{add}(\bar{\Delta})$ and $w_{IM}^{epi} = \widehat{w}_{IM}^{epi}(\bar{\Delta})$ (from Eqs. (4.32) and (4.33), respectively). The version of mBIC adapted to multiple regression interval mapping recommends choosing the model for which

$$mBIC_{IM} = n \log RSS + (k_1 + k_2) \log n + 2k_1 \log(w_{IM}^{add}N/2.2 - 1)$$
$$+ 2k_2 \log(w_{IM}^{epi}N_e/2.2 - 1),$$

obtains its minimal value.

In [7], versions of mBIC adapted to densely spaced markers and interval mapping are investigated on the basis of an extensive simulation study. This study shows that

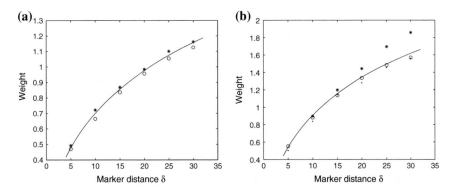

Fig. 4.7 Comparison of approximated and simulated weights for interval mapping. The circles and stars correspond to chromosomes of length 150 and 300 cM, respectively. **a** $\hat{w}_{SM}^{add}$, **b** $\hat{w}_{SM}^{int}$

they indeed control FWER in the weak sense at the assumed level. Moreover, it has been observed that densely spaced markers increase the power to detect QTLs and improve the accuracy of QTL localization. On the other hand, multiple interval mapping does not substantially increase power in comparison to single marker tests with loosely spaced markers. However, multiple interval mapping can substantially improve the accuracy of QTL localization.

4.4 Logic Regression

In the classical setting of generalized linear models described above, gene–gene interactions are understood in terms of the joint influence of several genes which cannot be explained by the sum of their additive effects. Therefore, interaction effects are often only considered in a model when the corresponding loci have main effects. This classical statistical definition of epistasis (interaction) goes back to Fisher [27] and is, in principle, different from its biological definition. Following Bateson and Mendel [5], biological epistasis is usually understood as a phenomenon in which "a variant or allele at one locus [...] prevents the variant or allele at another locus from manifesting its effect" [18], or more generally as a situation in which (for example) the effect of one allele can only be observed when a second allele is also present, and vice versa. A comprehensive comparison of these two definitions of epistasis can be found, for example, in [18] or [16]. Slightly extending the definition of Bateson and Mendel, biological epistasis occurs when a phenotype is affected by a specific combination of the genotypes of two or more genes. Consider, for example, a relatively simple situation in which a change in a phenotype (e.g., disease status) is caused by a mutation at locus A and locus B, or locus A but not locus B, or locus A and (locus B or locus C). As noted in [41], the identification of such an interaction using a classical approach based on gLMs becomes difficult, as it requires the consideration of complex linear models with a large number of parameters. Instead, such combinations

Table 4.2 Different codings of genotypes

Genotype of ith QTL	D_i	R_i	X_i	Z_i
$a_i a_i$	0	0	1	$-1/2$
$A_i a_i$	1	0	0	$1/2$
$A_i A_i$	1	1	-1	$-1/2$

of QTL genotypes can be directly coded in the form of logical expressions, which subsequently can be used as predictors in a linear model. This idea was used in [45], where the novel method of "logic regression" for identifying epistatic interactions in the form of logical expressions was proposed. In [43], this method was carefully studied in the context of QTL mapping in experimental populations.

To understand the concept of logic regression, consider the situation where the predictors $X_1, X_2, \ldots, X_m$ are binary. Any combination of these variables using the logical operators $\wedge$ (AND), $\vee$ (OR), and ^{C} (NOT) is called a logical expression (for example $L = (X_1 \wedge X_2) \vee X_3^C$). Any logical expression based on binary variables is also a binary variable. A logic regression model is a generalized linear model (gLM) of the form

$$g(E[Y]) = \beta_0 + \sum_{j=1}^{t} \beta_j L_j,$$

where $L_j, j = 1, 2, \ldots t$, are logical expressions, which can contain one or more primary binary predictors $X_p, p = 1, 2, \ldots, m$. In particular, if a logical expression contains only one such predictor, it corresponds to a main effect. The logical expressions L_j are called trees, and the primary variables contained in each tree are called leaves [38]. For example, the logical expression $L = (X_1 \wedge X_2) \vee X_3^C$ has three leaves: X_1, X_2, and X_3^C. The coefficients of a logic regression model can be estimated using classical maximum likelihood methods for GLMs.

In the context of QTL mapping in experimental populations, the genotypes are naturally binary for backcross and recombinant inbred lines (since there exist only two genotypes at a given locus). In the case of the intercross design, two binary predictors, denoted D_i and R_i, are needed for each locus. In Table 4.2, we compare these binary predictors with the classical state variables used in the standard Cockerham model X_i and Z_i. The regression coefficient of D_i (R_i) can be interpreted as the dominant (respectively, recessive) effect of allele A_i.

The basic difference between the linear approach used in the classical Cockerham model and the logic regression approach is illustrated by the following simple example. Consider the genotypes of m markers from an intercross $F2$ design. Assume that the phenotype Y is related to only two of these markers combined according to some binary logical rule. For example, suppose the true model is described by the formula

$$Y = \beta_0 + \beta_1 \cdot (D_k \wedge D_l) + \varepsilon, \qquad \varepsilon \sim \mathcal{N}(0, \sigma^2), \qquad (4.34)$$

which corresponds to the situation where the trait of interest is modified only in individuals who have two copies of the major allele at both of the loci k and l. We also assume that we do not know the true model, but only the fact that it involves one two-way interaction.

To identify the correct model using logic regression, the correct two-way logical expression has to be chosen from the set of all possible two-way logical expressions and the corresponding regression parameter needs to be estimated. To determine the total number of all possible two-way logical expressions, observe that each marker can be represented by four dummy variables: dominant and recessive, as well as their complements. Furthermore, the dummy variables corresponding to two markers can be combined using two logical operators: $\wedge$ (AND) and $\vee$ (OR). We do not need to consider the complements of the final logical expression, since the test for the significance of any logical predictor is equivalent to the test for the significance of its complement. Thus, the total number of two-component logical expressions is equal to $\binom{m}{2} \cdot 4^2 \cdot 2$. From de Morgan's Laws, it is sufficient to consider only AND combinations here. Hence, we can additionally divide this number by 2. Thus there are $\binom{m}{2} \cdot 4^2$ different two-component logical expressions. Hence, in this case, the model space which must be searched is determined by models of the form

$$M^{(j)} : \ Y = \tilde{\beta}_0^{(j)} + \tilde{\beta}_1^{(j)} \cdot L^{(j)} + \tilde{\varepsilon}^{(j)}, \qquad \tilde{\varepsilon}^{(j)} \sim \mathcal{N}(0, \sigma^2), \qquad (4.35)$$

where $j \in \mathscr{J} = \{1, 2, \ldots, \binom{m}{2} \cdot 4^2\}$, and $L^{(j)}$ can be any expression of the form $U_r \wedge U_s$, with $r, s \in \{1, 2, \ldots, m\}$, $r \neq s$ and $U_p \in \{D_p, R_p, D_p^C, R_p^C\}$ for $p \in \{r, s\}$.

On the other hand, to identify the correct model using the linear approach based on Cockerham's coding, one needs to consider all the possible linear models involving two markers. This means that, for each pair of variables, one needs to check a model containing the additive and dominant effect for each marker (4 parameters) and four different epistatic effects described by products of variables corresponding to the main effects of the two markers. Hence, one needs to estimate 8 parameters for each pair of markers. For the pair of markers from the true model (4.34), Cockerham's model describing the relationship between the phenotype Y and the two markers is of the form

$$M^{(k,l)} : \ Y = \mu_{(k,l)} + \beta_k X_k + \beta_l X_l + \delta_k Z_k + \delta_l Z_l + \gamma_{x_k x_l} X_k X_l + \gamma_{x_k z_l} X_k Z_l$$
$$+ \ \gamma_{z_k x_l} Z_k X_l + \gamma_{z_k z_l} Z_k Z_l + \varepsilon, \qquad \varepsilon \sim \mathcal{N}(0, \sigma^2). \qquad (4.36)$$

The total number of full linear models involving two markers is equal to the number of possible choices of two from m markers, $\binom{m}{2}$, and the model space to be searched through is $\{M^{(k,l)} : k, l \in \{1, 2, \ldots, m\}, k < l\}$. This simple example shows that

1. When using logic regression, one may represent the influence of several markers in a concise form, which, compared to the classical approach, reduces the number of parameters in the model. This increases the accuracy of the estimation of model parameters and the power to detect influential genes.

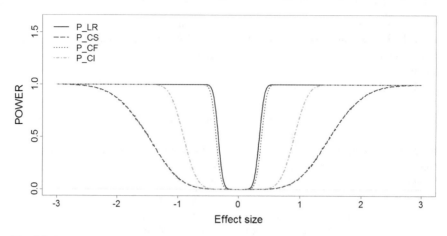

Fig. 4.8 Power to detect an interaction using various regression models as functions of β_1, for $n = 1000$, $m = 200$, $FWER = 0.05$, $\sigma^2 = 1.0$. P_{LR}—logic regression test for interaction, P_{CS}—simple regression test for interaction, based on the Cockerham coding, P_{CF}—full two-way ANOVA test, P_{CI}—test for a two-way interaction using the full Cockerham model

2. The number of possible logic regression models to be visited is considerably larger than the number of Cockerham's models. Hence, to prevent a large number of false discoveries using logic regression, a substantially stricter correction for multiple testing needs to be used [10], which potentially leads to a decrease in power.

   Figure 4.8 illustrates the results of calculations of the theoretical power to detect this interaction and a simulation study, explained in detail in [43]. It shows that, in our example, the advantage of using the simple logic regression model outweighs the cost of the loss in power resulting from correcting the effect of multiple testing. Even after using a more stringent correction for multiple testing, the power to detect two-way interactions using the logic regression model is slightly larger than the power to detect the influence of two genes using an F test appropriate for the full ANOVA model with interactions. Moreover, the power to detect this interaction within the selected two-way ANOVA model is much smaller than the corresponding power provided by logic regression. This results from the fact that this two-way logical interaction is also projected onto the classical main effects and the remaining component, corresponding to classical "interaction," is relatively weak. The results of extensive simulation studies and analysis of real data reported in [43] provide more examples of the desirable properties possessed by logic regression and suggest that it is a very valuable complement to classical methods of QTL mapping. In these simulations, interesting logical interactions were identified using the *logicFS* algorithm developed in [47] and used to build a logic regression model based on classical greedy step-wise techniques, adjusted for multiple testing.

4.5 Applying mBIC in a Bayesian Approach

Apart from the classical methods of QTL mapping discussed above, there exist a variety of Bayesian approaches, where the number of QTLs and other parameters of the genetic model are estimated using Markov chain Monte Carlo (MCMC) methods, discussed in Sect. 9.3. A partial list of references is provided in the interesting research manuscripts: [33, 57] or [60]. Bayesian approaches utilize prior knowledge. Moreover, the posterior distributions resulting from Bayesian analysis enable direct evaluation of the precision of statistical inference. However, one drawback of Bayesian reasoning is the rather large computational cost of such statistical inference. For example, when applying MCMC, the posterior distribution of a given statistical model is usually estimated by the frequency with which it appears in the Markov chain. This means that good models need to be visited many times. On the other hand, the modified versions of BIC discussed in this chapter can be utilized to approximate the posterior probability according to the formula

$$P(M_i|Y) \approx \frac{\exp(-xBIC(i)/2)}{\sum_j \exp(-xBIC(j)/2)}, \qquad (4.37)$$

where $xBIC$ denotes $mBIC$, $mBIC2$, or $EBIC$ and the sum in the denominator is over all the regression models visited. This estimator will be accurate if the sum in the denominator includes a representative number of "good" models. In [28], such a large population of models is obtained by applying an appropriately defined genetic algorithm. The simulations reported in [28] illustrate the very good statistical properties of such a method of QTL mapping, based on the estimated posterior probabilities of the presence of a QTL in a given neighborhood of a marker. Moreover, this genetic algorithm often finds a model with a better value according to mBIC than the best model found by commonly applied greedy step-wise methods.

4.6 Closing Remarks

Despite QTL mapping's long history, this field of research remains very active and new methods are constantly being developed. Therefore, we urge the interested reader to reach beyond the literature presented in this chapter. Among the most interesting research directions, one should mention the development of methods aimed at identifying genes with time-dependent effects, discussed in [30, 42]. In practical applications of QTL mapping, researchers can rely on a variety of statistical software, e.g., *QTL Cartographer*, or recent *R* libraries such as *R/qtl*, which has implementations of many different methods of QTL mapping and is described in [11], or *R/qtlbim*, focused on Bayesian QTL mapping. Also, we suggest careful inspection of updates, since new, freely available statistical programs with novel methods of QTL mapping appear regularly.

References

1. Allison, D.B., Fernandez, J.R., Heo, M., Zhu, S., Etzel, C., Beasley, T.M., Amos, C.I.: Bias in estimates of quantitative trait locus effect in genome scans: demonstration of the phenomenon and a methods-of-moments procedure for reducing bias. Am. J. Hum. Genet. **70**, 575–585 (2002)
2. Anderson, V.L., Kempthorne, O.: A model for the study of quantitative inheritance. Genetics **39**, 883–898 (1954)
3. Baierl, A., Bogdan, M., Frommlet, F., Futschik, A.: On locating multiple interacting quantitative trait loci in intercross designs. Genetics **173**, 1693–1703 (2006)
4. Baierl, A., Futschik, A., Bogdan, M., Biecek, P.: Locating multiple interacting quantitative trait loci using robust model selection. Comput. Stat. Anal. **51**, 6423–6434 (2007)
5. Bateson, W.: Mendels Principles of Heredity. Cambridge University Press (1902)
6. Bogdan, M., Ghosh, J.K., Doerge, R.W.: Modifying the Schwarz Bayesian information criterion to locate multiple interacting quantitive trait loci. Genetics **167**, 989–999 (2004)
7. Bogdan, M., Frommlet, F., Biecek, P., Cheng, R., Ghosh, J.K., Doerge, R.W.: Extending the modified Bayesian Information Criterion (mBIC) to dense markers and multiple interval mapping. Biometrics **64**, 1162–1169 (2008)
8. Bogdan, M., Żak-Szatkowska, M., Ghosh, J.K.: Selecting explanatory variables with the modified version of Bayesian information criterion. Qual. Reliab. Eng. Int. **24**, 627–641 (2008)
9. Bogdan, M., Doerge, R.W.: Biased estimators of QTL heritability and location in interval mapping. Heredity **95**, 476–484 (2005)
10. Boulesteix, A.-L., Strobl, C., Weidinger, S., Wichmann, H.-E., Wagenpfeil, S.: Multiple testing for SNP-SNP interactions. Stat. Appl. Gen. Mol. Biol. **37**, 6 (2007)
11. Broman, K.W., Sen, S.: A Guide to QTL Mapping with R/qtl. Springer, New York (2009)
12. Broman, K.W., Speed, T.P.: A model selection approach for the identification of quantitative trait loci in experimental crosses. J. R. Stat. Soc.: Series B (Stat. Meth.) **64**(4), 641–656 (2002)
13. Broman, K.W.: Review of statistical methods for QTL mapping in experimental crosses. Lab Anim. **30**(7), 44–52 (2001)
14. Broman, K.: Mapping quantitative trait loci in the case of a spike in the phenotype distribution. Genetics **163**(3), 1169–1175 (2003)
15. Churchill, G.A., Doerge, R.W.: Empirical threshold values for quantitative trait mapping. Genetics **138**, 963–971 (1994)
16. Clayton, D.G.: Prediction and interaction in complex disease genetics: experience in type 1 diabetes. PLoS Genet. **5**, e1000540 (2009)
17. Conneally, P.M., Edwards, J.H., Kidd, K.K., Lalouel, J.-M., Morton, N.E., et al.: Report of the committee on methods of linkage analysis and reporting. Cytogenet. Cell Genet. **40**, 356–359 (1985)
18. Cordell, H.J.: Epistasis: what it means, what it doesnt mean, and statistical methods to detect it in humans. Hum. Mol. Genet. **11**, 2463–2468 (2002)
19. Cui, Y., Kim, D.-Y., Zhu, J.: On the generalized Poisson regression mixture model for mapping quantitative trait loci with count data. Genetics **3**, 2159–2172 (2006)
20. Cui, Y., Yang, W.: Zero inflated generalized Poisson regression mixture model for mapping quantitative trait loci underlying count trait with many zeros. J. Theor. Biol. **256**(2), 276–285 (2009)
21. Dempster, A.P., Laird, N.M., Rubin, D.B.: Maximum likelihood from incomplete data via EM algorithm. J. Roy. Stat. Soc. Ser. B **39**, 1–38 (1977)
22. Dupuis, J., Siegmund, D.O.: Statistical methods for mapping quantitative trait loci from a dense set of markers. Genetics **151**, 373–386 (1999)
23. East, E.M.: Studies on size inheritance in Nicotiana. Genetics **1**, 164–176 (1916)
24. Emerson, R.A., East, E.M.: The inheritance of quantitative characters in maize. Nebr. Agric. Exp. Stat. Res. Bull. No. 2: 1–120 (1913)

25. Erhardt, V., Bogdan, M., Czado, C.: Locating multiple interacting quantitative trait loci with the zero-inflated generalized Poisson regression. Stat. Appl. Genet. Mol. Biol. **9**(1), Article 26 (2010)
26. Feingold, E., Brown, P.O., Siegmund, D.: Gaussian models for genetic linkage analysis using complete high resolution maps of identity-by-descent. Am. J. Hum. Genet. **53**, 234–251 (1993)
27. Fisher, R.A.: The correlation between relatives on the supposition of Mendelian inheritance. Philos. Trans. R. Soc. Edinb. **52**, 399–433 (1918)
28. Frommlet, F., Ljubic, I., Arnardottir, H., Bogdan, M.: QTL Mapping using a memetic algorithm with modifications of BIC as fitness function. Stat. Appl. Genet. Mol. Biol. **11**(4), Article 2 (2012)
29. Göring, H.H.H., Terwilliger, J.D., Blangero, J.: Large upward bias in estimation of locus-specific effects from genomewide scans. Am. J. Hum. Genet. **69**, 1357–1369 (2001)
30. Hadjipavlou, G., Bishop, S.C.: Age-dependent quantitative trait loci affecting growth traits in scottish blackface sheep. Anim. Genet. **40**, 165–175 (2009)
31. Haley, C.S., Knott, S.A.: A simple regression method for mapping quantitative trait loci in line crosses using flanking markers. Heredity **69**, 315–324 (1992)
32. Hayes, H.K., East, E.M.: Further experiments on inheritance in maize. Conn. Agr. Exp. Stat. Bull. **188**, 1–31 (1915)
33. Huang, H., Eversley, C.D., Threadgill, D.W., Zou, F.: Bayesian multiple quantitative trait loci mapping for complex traits using markers of the entire genome. Genetics **176**, 2529–2540 (2007)
34. Huber, P.J.: Robust Statistics. Wiley, New York (1981)
35. Jurečková, J., Sen, P.K.: Robust Statistical Procedures: Asymptotics and Interrelations. Wiley, New York (1996)
36. Kao, C., Zeng, Z., Teasdale, R.: Multiple interval mapping for quantitative trait loci. Genetics **152**, 1203–1216 (1999)
37. Kao, C.: On the differences between maximum likelihood and regression interval mapping in the analysis of quantitative trait loci. Genetics **156**, 855–865 (2000)
38. Kooperberg, C., Ruczinski, I.: Identifying interacting SNPs using Monte Carlo logic regression. Genet. Epidemiol. **28**, 157–170 (2005)
39. Kruglyak, L., Lander, E.S.: A nonparametric approach for mapping quantitative trait loci. Genetics **139**, 1421–1428 (1995)
40. Lander, E.S., Botstein, D.: Mapping Mendelian factors underlying quantitative traits using RFLP linkage maps. Genetics **121**, 185–199 (1989)
41. Lucek, P.R., Ott, J.: Neural network analysis of complex traits. Genet. Epidemiol. **14**, 1101–1106 (1997)
42. Lund, M.S., Sorensen, P., Madsen, P., Jaffrezic, F.: Detection and modelling of time-dependent QTL in animal populations. Genet. Sel. Evol. **40**, 177–194 (2008)
43. Malina, M., Ickstadt, K., Schwender, H., Posch, M., Bogdan, M.: Detection of epistatic effects with logic regression and a classical linear regression model. Stat. Appl. Genet. Mol. Biol. **13**, 83–104 (2014)
44. Rebaï, A., Goffinet, B., Mangin, B.: Approximate thresholds of interval mapping test for QTL detection. Genetics **138**, 235–240 (1994)
45. Ruczinski, I., Kooperberg, C., LeBlanc, M.: Logic regression. J. Comput. Graph. Stat. **12**, 474–511 (2003)
46. Sax, K.: The association of size differences with seed-coat pattern and pigmentation in Phaseolus vulgaris. Genetics **8**, 552–560 (1923)
47. Schwender, H., Ickstadt, K.: Identification of SNP interactions using logic regression. Biostatistics **9**, 187–198 (2008)
48. Siegmund, D., Yakir, B.: Significance level in interval mapping. In: Zhang, H., Huang, J. (eds.) Development of modern statistics and related topics in celebration of Yaoting Zhangs 70th birthday. World Scientific, Singapore (2003)
49. Siegmund, D., Yakir, B.: The Statistics of Gene Mapping. Springer, Springer Series in Statistics for Biology and Health (2007)

50. Soller, M., Brody, T., Genizi, A.: On the power of experimental designs for the detection of linkage between marker loci and quantitative loci in crosses between inbred lines. Theor. Appl. Genet. **47**, 35–39 (1976)
51. Stewart, J.: Biometrical genetics with one or two loci: I. The choice of a specific genetic model. Heredity **24**, 211–224 (1969a)
52. Stewart, J.: Biometrical genetics with one or two loci: II. The estimation of linkage. Heredity **24**, 225–238 (1969b)
53. Stewart, J., Elston, R.C.: Biometrical genetics with one or two loci: the inheritance of physiological characters in mice. Genetics **73**, 675–693 (1973)
54. Thoday, J.M.: Location of polygenes. Nature **191**, 368–370 (1961)
55. Xu, S., Atchley, W. R.: Mapping quantitative trait loci for complex binary diseases using line crosses. Genetics, **143**(3), 1417–24 (1996)
56. Xu, S.: Theoretical basis of the Beavis effect. Genetics **165**, 2259–2268 (2003)
57. Yi, N., Yandell, B.S., Churchill, G.A., Allison, D.B., Eisen, E.I., Pomp, D.: Bayesian model selection for genome-wide epistatic QTL analysis. Genetics **170**, 1333–1344 (2005)
58. Żak, M., Baierl, A., Futschik, A., Bogdan, M.: Locating multiple interacting quantitative trait loci using rank-based model selection. Genetics **176**, 1845–1854 (2007)
59. Żak-Szatkowska, M., Bogdan, M.: Modified versions of Bayesian information criterion for sparse generalized linear models. Comput. Stat. Data Anal. **55**, 2908–2924 (2008)
60. Zou, F., Huang, H., Lee, S., Hoeschele, I.: Nonparametric Bayesian variable selection with applications to multiple quantitative trait loci mapping with epistasis and geneenvironment interaction. Genetics **186**, 385–394 (2010)

Chapter 5
Statistical Analysis of GWAS

5.1 Overview

As described previously (see Sect. 2.2.3), the main difference between QTL mapping and association studies is the type of population from which genetic data are collected. While QTL mapping is based on experimental populations, association studies usually deal with natural populations. Association studies have a long history, and the excellent review article of Balding [7] gives an introduction to their most important statistical aspects. Today, most association studies use single nucleotide polymorphisms (SNPs) as markers. Thanks to efforts like the HapMap project [52] and the 1000 genomes project [123], the location of more than 10 million SNPs within the human genome are now known.

Genome-wide association studies (GWAS) have become feasible with the development of SNP array technology, as described in [63]. Affymetrix havef developed SNP arrays for an ever-increasing number of SNPs, starting with 10, 100, and 500 K arrays, followed by SNP Array 5.0 and SNP Array 6.0. Illumina (Bead-Chips) and Roche (NimbleGen arrays) are competing producers of SNP arrays. The Affymetrix Genome-Wide Human SNP Array 6.0 covers more than 900,000 SNPs, and additionally more than 900,000 non-polymorphic probes for detecting copy-number variants (CNVs). The HumanOmni5-Quad BeadChip comprises approximately 4.3 million markers (among them more than 2 million SNPs) and it is possible to add some 500,000 custom-made polymorphisms. In our presentation of statistical issues, we will mainly focus on SNPs, though many of these considerations will also apply to CNVs.

The first major GWAS were published in 2007, and since then hundreds of GWAS have been performed. Their number is rapidly increasing, as documented by the regularly updated catalog of published genome-wide association studies [56] from the documents of the National Human Genome Research Institute. Figure 5.1 is based on this catalog and illustrates the success of GWAS in the period 2007–2010.

© Springer-Verlag London 2016
F. Frommlet et al., *Phenotypes and Genotypes*, Computational Biology 18,
DOI 10.1007/978-1-4471-5310-8_5

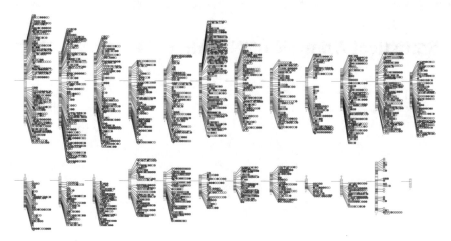

Fig. 5.1 Regions of the human genome where associations have been detected by published GWAS studies as of January 2016. The diagram is taken from http://www.ebi.ac.uk/gwas/diagram

In 2005, even before any GWAS had been carried out, Hirschhorn and Daly [57] gave a concise summary of the problems involved in them. They gave an argument for why GWAS are useful and discussed the inherent multiple testing problem in depth, along with strategies of dealing with it. Several other reviews dealing with the statistical aspects involved in GWAS are available, see for example [20, 79, 134].

The first task of GWAS is to infer the genotype of SNPs from the DNA array image data. A brief and non-exhaustive survey of methods for inferring genotypes (genotype calling) is given in Sect. 5.2.1. Often, genotype calling for a specific SNP is not successful for all individuals, but one would still like to use such SNPs in an analysis. Also, one might be interested in the genotypes at sites which have not been previously genotyped, and sometimes one is actually interested in inferring not only genotypes, but haplotypes. Strategies for imputing genotypes and haplotype phasing are discussed in Sect. 5.2.2.

After genotypes have been called, the actual association analysis is performed. In Sect. 5.3, the most common statistical approaches based on single markers are reviewed. In terms of a frequentist approach to the analysis of GWAS, it has become standard to start by computing p-values from single marker tests and then consider markers associated with realizations of a test statistic above some threshold reflecting correction for multiple testing [40]. However, as pointed out for example in [57], genome-wide association studies are based on the assumption that common diseases are associated with common variants. This hypothesis suggests that common diseases result from a moderate number of small common genetic effects. Given this hypothesis is true, such single marker tests are bound to fail, as was explained exhaustively in [48]. Recently, a number of approaches have been introduced to analyze GWAS using strategies for model selection, which will be presented in Sect. 5.4.

5.2   Inferring Genotypes

SNP arrays make use of the principle of hybridization, which means that complementary strands of nucleic acid will bind to each other. Based on microarray technology, small pieces of DNA (probes) can be fixed onto the surface of an array. Since the distance between probes is only a few micrometers, it is possible to have several million probes on a single array. To determine the genotype of a certain SNP, different kinds of probes are fixed on the array, some of which correspond to the reference (major) allele a and others to the variant (minor allele) A at the relevant position. The aim is to measure how much of the prepared DNA in a sample (target) is hybridized at each probe. If the target only binds at the a probe (or only at the A probe), one concludes that the sample is homozygous aa (or AA) at that locus. Binding at both probes indicates heterozygosity, Aa.

To quantify the amount of hybridization, targets are usually labeled with certain chemicals. Specialized scanners can measure the abundance of these chemicals at each probe. Therefore, the raw data for GWAS are image data which represent the intensities of the signals measured by the scanners. In practice, one is rarely confronted with the raw image data, but the various GWAS platforms provide summary intensity scores for each probe. Affymetrix arrays provide intensities for each probe in .cel files. The following section considers how to infer genotypes based on these data. We focus particularly on methods developed for Affymetrix arrays. For genotype calling based on data from Illumina BeadChips, see [106] and the references given there.

5.2.1   Genotype Calling

The procedure for inferring genotypes from the probe intensities corresponding to the two alleles at a particular locus is referred to as genotype calling. Before we discuss specific algorithms in detail, look at Fig. 5.2, which illustrates the problem to be solved. Scatter plots of normalized probe intensities are shown for two different SNPs. Each point corresponds to one of the 90 CEU samples from HapMap (compare with Sect. 2.2.3 for more details on this sample), where the .cel data files from Affy500K arrays were downloaded from http://hapmap.ncbi.nlm.nih.gov/downloads/raw_data/affy500k/. Normalization and genotype calling were performed using the Bioconductor Oligo package [23].

Plot (a) illustrates an ideal situation. One observes three clearly distinct clusters corresponding to the three possible genotypes AA (black stars), Aa (red circles), and aa (blue crosses), and the calling algorithm has no problems in identifying the correct genotypes. In plot (b), the situation is rather simple. However, the minor allele frequency (MAF) is rather small, and there is only one homozygous aa genotype among the 90 individuals. For most SNPs, genotype calling is rather unambiguous and different algorithms lead to the same results. However, a small percentage of

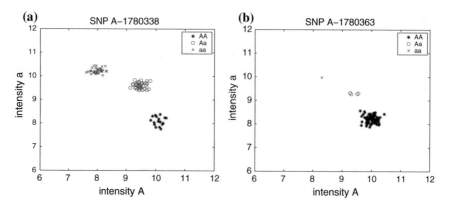

Fig. 5.2 Scatterplots of normalized intensities for SNP variants. Plots **a** and **b** show typical examples where genotyping works perfectly well

SNPs pose a greater challenge, and the performance of the algorithm applied is then crucial to ensuring correct genotyping. Other types of problems in genotyping which cannot be resolved by using better clustering algorithms, but are due to certain genetic constellations, are discussed, for example, in [134].

Genotype calling has two essential ingredients: the normalization of probe intensities and classification of genotypes based on these normalized intensities. Over the last few years, a large number of calling algorithms have been developed. A nice review of the early developments is given by [22]. SNP array platforms have their own inbuilt genotyping algorithms. For the earliest 10 K chips, Affymetrix used an algorithm called MPAM (Modified Partitioning Around Medioids, [73]) and then later developed DM (Dynamic Model Mapping Algorithm [38]) for the 100 K arrays. In response to certain shortcomings of these algorithms, Rabee and Speed [103] introduced RLMM (Robust Linear Model with Mahalanobis Distance Classifier). RLMM first applies quantile normalization and then fits an RMA model which is now a well-established approach to analyzing gene expression microarrays. In this multichip approach, the parameters of the model are estimated for each SNP from the 269 individuals genotyped by the HapMap project, a training set of data for which all the genotypes at all the SNPs are available. A Gaussian mixture model is then applied to classify the genotypes.

Next, Affymetrix implemented BRLMM (Bayesian RLMM) to call genotypes based on their 500 K Array. This algorithm does not train on the HapMap data, but instead uses DM to get a preliminary classification, which is then modified via a Bayesian approach. In [22], BRLMM is compared to CRLMM (Corrected RLMM), a refinement of RLMM which uses more complex models taking into account additional factors which influence intensity measurements. In particular, CRLMM considers fragment length and sequence effects, as well as differences between laboratories. For each SNP, Affymetrix arrays have probes from both strands of the DNA (sense and antisense). CRLMM fits models for these sense and antisense

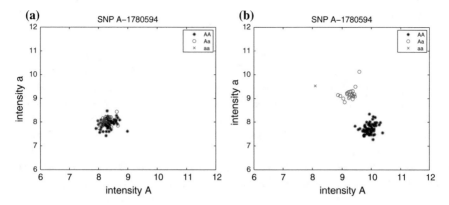

Fig. 5.3 Scatterplots of normalized intensities for SNP variants. Plots **a** and **b** show data for the sense and antisense strand of the same SNP

probes separately, which for a certain number of SNPs increases the ability to classify genotypes correctly. SNP A-1780594 is given as an example in Fig. 5.3. In the first plot, normalized intensities are shown for probes from the sense strand, based on which genotyping would be entirely impossible. On the other hand, the clusters corresponding to the antisense strand, given in the second plot, are rather clear. It is argued in [22] that developing separate models for the two strands increases the ability to genotype correctly. In general, CRLMM was shown to outperform BRLMM. CRLMM is now available in the Bioconductor package Oligo [23], which appears to have become the most popular open source algorithm for genotyping.

For their latest generation SNP Array 6.0, Affymetrix developed the birdseeds algorithm, which uses an EM (Expectation Maximization) algorithm to fit a model of a two-dimensional mixture of Gaussian distributions to normalized intensities.

Whichever algorithm one uses for genotype calling, one important issue is the question of quality control. It is clear that certain situations are difficult to resolve, whatever algorithm is used. For example, if one is confronted with a scatter plot like the one in Fig. 5.3a, but for both strands. A number of problematic cases are nicely illustrated in [111]. In principle, visual inspection can often help to resolve difficulties in genotype calling, but due to the huge number of SNPs to be called, visual inspection of all the images is simply not feasible. Therefore, there is a need for automatic procedures, and many genotype calling algorithms provide some measure of reliability, according to which one can judge whether problems in determining a specific genotype may have occurred. Visual inspection is also recommended for any SNPs which are found to be associated with the trait in question.

5.2.2   Imputation

When genotype calling algorithms are not able to unambiguously decide upon the genotype of a SNP, this leads to a missing value. For a given SNP, it might happen that genotype calling does not work well in general, i.e., for a large set of individuals. In such a case, one usually discards that SNP from further analysis. On the other hand, problems with a SNP array can lead to genotype calls of particularly poor quality for a particular individual, and one might have to exclude the data from that individual. Otherwise, if genotype data are only missing sporadically, one can use methods of imputation to call genotypes. In particular, in view of the multilocus models we will consider in Sect. 5.4, it is important to establish the genotypes for all SNPs. Comprehensive reviews on the methods of imputation for GWAS are given in [17, 78].

To impute missing genotypes, it is often sufficient to use SNP data from the study itself. Methods which use reference panels will be discussed below. An extremely simple heuristic strategy for imputation was suggested by [48]. To infer the genotype of a SNP with missing values, the information from strongly correlated SNPs within a neighborhood was used (when available). This type of heuristic imputation has the advantage of being fast, and when only a small percentage of data are missing the inference regarding unknown genotypes is still quite reliable.

More sophisticated methods of imputation are often based on haplotypes, and then the problem of inferring missing genotype data involves phasing, i.e., first inferring haplotypes from genotype data (see also Sect. 2.2.3). Perhaps the most popular program for phasing is FastPHASE [113]. Beagle [19] is a widely used alternative. Both programs use hidden Markov models to infer haplotypes from genotype data, where the differences between the underlying models are nicely explained in [17].

Methods of imputation can also be used in GWAS in a way which goes beyond the inference of sporadically missing data. There exist strategies for inferring the genotype of SNPs which have not been directly read. The purpose of this approach is similar to the rationale behind interval mapping for experimental populations, where inference based on the location of potential markers between real markers can increase power (see Chap. 4). As we have seen in Sect. 2.2.3, the correlation structure in outbred populations is much more complicated than in experimental populations, which makes the imputation of the genotypes of unread SNPs a much more challenging task. The main tool is to use information from reference panels, for example haplotype data from HapMap or from the 1000 genome project. An in-depth discussion of methods of imputation based on reference panels is given by [78].

All the statistical methods discussed below assume that missing data have been imputed. In particular, the methods of model selection presented in Sect. 5.4 rely upon complete data.

5.3 Single Marker Tests

The huge number of SNPs which have to be analyzed in GWAS is challenging, from both a statistical and computational point of view. Note that the number of markers in GWAS is often several orders larger than in QTL mapping. This is probably the main reason that carrying out a sequence of single marker tests independently of each other is still the most common practice in analyzing GWAS, although we will describe some severe shortcomings of this approach in Sect. 5.4.1. Nevertheless, we will start by introducing the single marker tests most commonly used in association studies. An excellent review on the statistical issues involved in the general field of association studies is given by Balding [7], whereas [134] focuses on GWAS.

5.3.1 Case-Control Studies

In Sect. 2.2.3.1, various designs of association studies were discussed, among which the case-control design is the most common in practice. Considering a single SNP, the data can be summarized in a contingency table, as in Table 5.1, where we denote the number of cases with genotype g by r_g and the number of controls with genotype g by s_g, where $g \in \{AA, Aa, aa\}$. The remaining notation denotes the marginal frequencies in the contingency table (the appropriate row and column sums). R and S are the total number of cases and controls, respectively (the column sums). The n_g denote the total number of individuals with genotype g (the row sums) and N is the total number of individuals.

In principle, one is interested in whether there is any association at all between the genotype of a SNP and the presence or absence of a disease. There are two statistical tests commonly used to answer this research question, Pearson's chi-square test for independence (with two degrees of freedom) or Fisher's exact test. Both tests are described in detail in Sect. 8.8 of the Appendix.

Pearson's chi-square and Fisher's exact test can be thought of as omnibus tests, which are able to detect any possible genetic association. In Fig. 5.4, we have four different situations which might occur in practice. In panel (a), for all three genotypes the proportion of individuals who are cases is roughly equal, and thus one might conclude that this SNP is not associated with the disease status. To be more precise, one can only conclude that the SNP has no marginal effect. It might still be the case

Table 5.1 Contingency table of genotype frequencies for cases and controls

Cases	Controls	Total
r_{AA}	s_{AA}	n_{AA}
r_{Aa}	s_{Aa}	n_{Aa}
r_{aa}	s_{aa}	n_{aa}
R	S	N

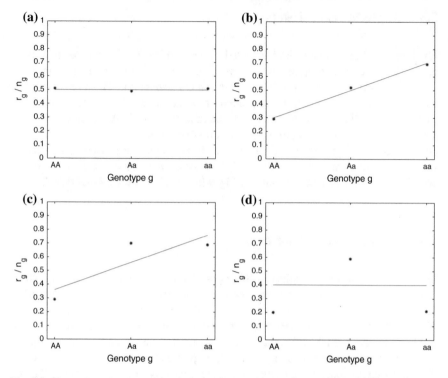

Fig. 5.4 Illustration of how suitable the Cochran–Armitage trend test (CAT) is for different genetic models. The *dots* represent the proportion of individuals with the given genotype who are cases. The *blue line* stems from least squares regression when genotypes are coded by equally spaced numbers (color online)

that the SNP has some effect on the disease status in combination with other SNPs, but, by definition, single marker tests do not enable us to make inferences of this kind.

Panel (b) of Fig. 5.4 shows the common situation of when a so called additive effect exists. The relative risk r_{Aa}/n_{Aa} for a heterozygous individual is larger than the relative risk r_{AA}/n_{AA} for a homozygous AA individual. On the other hand, a homozygous aa individual has an even larger risk r_{aa}/n_{aa}. In this case, a is called the risk allele. Given that just an additive effect exists, it follows that

$$\frac{r_{Aa}}{n_{Aa}} - \frac{r_{AA}}{n_{AA}} = \frac{r_{aa}}{n_{aa}} - \frac{r_{Aa}}{n_{Aa}},$$

i.e., the relative risk increases linearly in the number of a alleles in the genotype. Pearson's chi-square and Fisher's exact test are able to detect additive effects, but if one knows in advance that an additive model holds, then there exist more powerful tests.

In particular, the **Cochran–Armitage trend test** (CAT) is designed to detect additive effects [6]. One way to carry out the CAT is to encode the genotype states just as in Sect. 2.2.2 with $X = -1$ and $X = 1$ denoting the homozygote genotypes and $X = 0$ denoting heterozygotes, while the dichotomous trait is encoded as $Y = 1$ for cases and $Y = 0$ for controls. CAT is then equivalent to performing linear regression and carrying out an approximate z-test for the slope of the regression line. More specifically, least squares regression gives the slope

$$b_1 = \frac{N(r_{Aa} + 2r_{aa}) - R(n_{Aa} + 2n_{aa})}{N(n_{Aa} + 4n_{aa}) - (n_{Aa} + 2n_{aa})^2}$$

and the CAT statistic is defined as

$$z_C^2 := \frac{N b_1^2 \operatorname{Var}(X)}{\operatorname{Var}(Y)} = \frac{N\left[N(r_{Aa} + 2r_{aa}) - R(n_{Aa} + 2n_{aa})\right]^2}{R(N - R)\left[N(n_{Aa} + 4n_{aa}) - (n_{Aa} + 2n_{aa})^2\right]} . \tag{5.1}$$

Under the null hypothesis of no association, the CAT statistic is approximately chi-square distributed with one degree of freedom, $z_C^2 \sim \chi_1^2$.

If the additive model holds, then CAT will be more powerful than Pearson's chi-square or Fisher's exact test. However, associations do not always follow the additive model. Figure 5.4c illustrates the situation where the risk allele a is dominant, which is reflected in the fact that $\frac{r_{aa}}{n_{aa}} = \frac{r_{Aa}}{n_{Aa}} > \frac{r_{AA}}{n_{AA}}$. In such a situation, the CAT is no longer the most powerful test. The CAT works worst in the case of overdominance, depicted in Fig. 5.4d, where the CAT actually loses all its power, whereas the omnibus tests are still able to detect such an association.

The dominance model, illustrated in Fig. 5.4c, is of particular genetic relevance, just like the recessive model, which is the counterpart of the dominance model where a and A switch roles. This means that under the recessive model only individuals with two risk alleles aa are more susceptible to a disease, whereas AA and Aa individuals have the same risk. One can easily design Cochran–Armitage tests which are specifically suitable for the dominance or the recessive model, respectively. Consider the trend function

$$U = \sum_{\ell \in \{AA, Aa, aa\}} x_\ell (Sr_\ell - Rs_\ell),$$

where the scores x_ℓ define the trend one is interested in. In general, under the null hypothesis H_0 of no association, $E(U) = 0$. The CAT statistic is thus defined as $\frac{U^2}{\operatorname{Var}(U)}$, which under H_0 is asymptotically chi-square distributed with one degree of freedom. Computing $\operatorname{Var}(U)$ under the null hypothesis, one can show that using the scores $x_{AA} = -1$, $x_{Aa} = 0$ and $x_{aa} = 1$, the standard CAT statistic for detecting additive effects is obtained as $z_C^2 = \frac{U^2}{\operatorname{Var}(U)}$. In fact, any kind of linear scoring will lead to the standard CAT, whereas other scorings can be used to test other genetic models.

Table 5.2 Contingency table of allelic frequencies for cases and controls

	Cases	Controls	Total
Allel A:	$2r_{AA} + r_{Aa}$	$2s_{AA} + s_{Aa}$	$2n_{AA} + n_{Aa}$
Allel a:	$2r_{aa} + r_{Aa}$	$2s_{aa} + s_{Aa}$	$2n_{aa} + n_{Aa}$
	R	S	N

For example, to test the dominance model, one can use the scores $x_{AA} = -1$ and $x_{Aa} = x_{aa} = 1$. If the dominance model actually holds, then the resulting CA test will have maximal power. However, in practice, it is a priori unclear which genetic model is the correct one. For this reason, [44] suggested considering the maximum of the three CAT statistics for the additive, the dominance and the recessive model. We will discuss the properties of this max test in more detail in the next section in the context of quantitative traits.

Our main focus here is on tests based on genotype data, as illustrated in Table 5.1. In the genetics literature, one frequently comes across tests of association which instead make use of the allelic frequencies, which are accordingly illustrated in Table 5.2.

One might consider using chi-square tests for association based on the contingency table of observed allele frequencies. However, it was shown in [112] that this approach is only appropriate under Hardy–Weinberg equilibrium.

Hardy–Weinberg equilibrium (HWE):

Denote the relative genotype frequencies as $f(AA)$, $f(Aa)$ and $f(aa)$, and the relative allele frequencies as $p(A)$ and $p(a)$. A genetic locus is in Hardy–Weinberg equilibrium if

$$f(AA) = p(A)^2, \; f(Aa) = 2p(A)p(a), \; f(aa) = p(a)^2. \quad (5.2)$$

Furthermore, [112] also proved that under HWE chi-square tests based on allelic data are asymptotically equivalent to CA tests. Therefore, [112] gives a general recommendation to use tests based on genotypes.

One can test for a deviation from HWE using Pearson's goodness of fit test (see Sect. 8.8.1). Using (5.2), one computes the expected number of observations for each genotype under HWE as follows:

$$E_{AA} = \frac{(2n_{AA} + n_{Aa})^2}{4N}, \quad E_{Aa} = \frac{(2n_{AA} + n_{Aa})(2n_{aa} + n_{Aa})}{2N}, \quad E_{aa} = \frac{(2n_{aa} + n_{Aa})^2}{4N},$$

and Pearson's test statistic becomes

$$\chi^2 = \sum_{\ell \in \{AA, Aa, aa\}} \frac{(n_\ell - E_\ell)^2}{E_\ell}.$$

The test statistic is chi-square distributed with one degree of freedom, which relates to the difference between the number of genotypes (three) and the number of alleles (two). There also exists a version of Fisher's exact test for HWE.

In GWAS analysis, HWE is more or less routinely used as a quality criterion, where markers which strongly deviate from HWE are discarded. The rationale behind this it that in most cases deviation from HWE is due to artifacts of the genotyping process, see [68]. However, Balding argued in his tutorial [7] that in some cases deviation from HWE might also be related to true association.

5.3.2 Quantitative Traits

The simplest approach to analyzing the genetic association of a SNP with a quantitative trait is to make use of a general linear model (GLM). A brief introduction to GLMs is given in Sect. 8.4 of the Appendix, where it is pointed out that both analysis of variance (ANOVA), as well as regression are special cases of GLMs. There are certain analogies between the single marker tests we discussed above for case-control studies and the tests available for quantitative traits. For example, applying a one-way ANOVA resembles Pearson's χ^2 test for case-control studies, in the sense that such tests look for any sort of differences between the mean values of traits for the three genotypes. The corresponding F-test statistic for the ANOVA again has two degrees of freedom.

If one assumes in advance that an additive model holds, then one can immediately test for a linear trend using a simple regression model, which corresponds to the Cochran–Armitage trend test for binary traits. Under a recessive or a dominance model, the data from the genotypes for which the trait value is expected to have the same mean can be pooled, and then one can apply a two-sample t-test to test for a difference between the means for the two remaining groups.

We want to point out that the term dominance effect can have different meanings depending on the context. As defined in Sect. 2.2.2, the dominance effect in the context of QTL mapping is usually used to describe the difference between the mean for heterozygotes and the average of the means for homozygotes. Let us look at this in the language of GLMs. If the genotype of a marker is encoded as $G \in \{-1, 0, 1\}$, then the additive effect is obtained using G as a regressor, whereas the dominance effect corresponds to G^2. Based on such a coding, the regressors G and G^2 correspond to the two degrees of freedom in the one-way ANOVA. If there is the same number of homozygotes for both alleles (as in an ideal F2 population in QTL mapping), then the regressors G and G^2 are orthogonal. This is particularly valuable from a statistical point of view, because then the effects of G and G^2 can be estimated independently.

In human genetics, a dominance effect usually refers to the situation depicted in Fig. 5.4c for binary traits, where those with genotypes Aa and aa have on average the same trait value, whereas the mean trait of those with genotype AA is different. One might say that allele a dominates allele A. In terms of a GLM, this corresponds to a regression model with a regressor of the form $D \in \{-1, 1, 1\}$. For the recessive model, one might consider regressors of the form $R \in \{-1, -1, 1\}$.

As mentioned above, it is not usually a priori known which genetic model is correct. One possible strategy is to consider the three most commonly used genetic models described above (represented by G, D and R) and then use the one with the largest realization of the test statistic (or, equivalently, the one which gives the smallest p-value). However, this strategy makes use of three different tests, and, as we have seen in Sect. 3.2, it is therefore necessary to correct for multiple testing. In [59], the maximum of the three test statistics corresponding to these models (the MAX test) was considered in the context of case-control studies, and a permutation test approach was explored to correct for multiple testing. Here, we want to illustrate the MAX test applied to quantitative traits, and we specifically want to compare its performance with the one-way ANOVA approach.

To this end, we performed a small simulation study based on 5000 replicates. We considered 500 heterozygotes and 250 homozygotes of each type. A quantitative trait was generated according to the linear model $Y_i = 0.1 \cdot X_i + \epsilon_i$, where $\epsilon_i \sim \mathcal{N}(0, 1)$ i.i.d. and $i \in \{1, \ldots, 1000\}$. We considered four different scenarios, which are specified in Table 5.3. In each scenario, we used a different value of the regressor X_i for heterozygotes. Scenario 1 models a purely additive effect, Scenario 3 a purely dominant effect, and Scenario 2 is somewhere in between. Finally, Scenario 4 considers the case of overdominance (compare with Fig. 5.4d).

Figure 5.5 depicts the results of this simulation study in the form of the empirical distribution functions of p-values obtained from the different tests. ANOVA refers to the F-test with two degrees of freedom. Additive, Dominance and Recessive refer to the t-tests (or equivalently F-tests with one degree of freedom) for regression based on these particular genetic models, respectively. Figure 5.5 also shows the distribution of $p_{\min}$, the minimal p-value from these three tests. The corresponding testing procedure is not adjusted for multiplicity, which makes direct comparison with the other tests impossible.

To calculate adjusted p-values for the MAX test, we first simulated 10,000 instances under the null hypothesis (i.e., $Y = \epsilon$ with $\epsilon \sim \mathcal{N}(0, 1)$) and computed the minimal p-value for the tests based on the Additive, Dominance, and Recessive models. This yields the (empirical) distribution F_0 of minimal p-values under the null hypothesis. Given a minimal p-value $p_{\min}$ under the alternative, we obtain the

Table 5.3 Four different simulation scenarios to compare the MAX test with the ANOVA approach

Scenario	AA	Aa	aa
1 (additive)	-1	0	1
2 (add + dom)	-1	0.5	1
3 (dominant)	-1	1	1
4 (overdominant)	-1	1	0

The table gives the values of X_i depending on the genotype of the ith individual

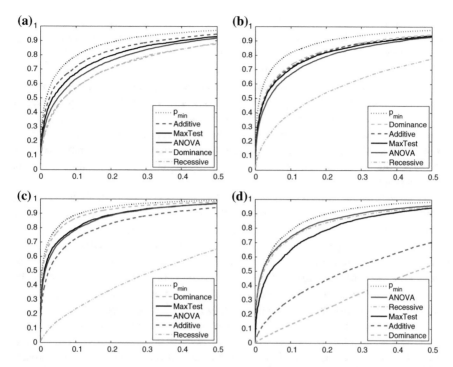

Fig. 5.5 The empirical distribution of p-values allows us to assess the performance of different strategies for single marker tests. Except for p_{min} all the tests control the type I error rate, and thus the graphs can be interpreted as showing the power for a given significance level

corresponding adjusted p-value by computing the probability of observing a minimal p-value which is the same or even smaller under the null hypothesis, i.e., $F_0(p_{min})$. We can apply this procedure, because in the simulation setting we can control the distribution of the error term, and thus know the underlying null distribution. In practice, a permutation test approach would lead to rather similar conclusions.

In the purely additive case (Scenario 1), it does not come as a surprise that the test based on the additive model alone is the most powerful. However, being uncertain about the underlying genetic model, one would do better to use the MAX test compared to the one-way ANOVA procedure. Similarly, in Scenario 2 the MAX test has greater power than the ANOVA procedure. Here, the MAX test is about as powerful as the test based on the additive model alone, whereas the test based on the dominance model is even more powerful. The test based on the dominance model is clearly the most powerful in Scenario 3, where the MAX test and ANOVA appear to perform similarly, with the MAX test only having a slight advantage. One-way ANOVA is more powerful than the other tests only in the case of overdominance (Scenario 4).

In summary, one might conclude that as long as the possibility of overdominance can be ruled out, but otherwise there is uncertainty about the underlying genetic model, then the MAX test can be recommended. It works rather well in each of the first three scenarios, without losing too much power compared to the optimal tests for Scenario 1 (additive model) and Scenario 3 (dominance model). Isotonic regression (see [10, 109]) might be even slightly more powerful for scenarios in which additive and dominance effects are combined. However, one would expect that the gain in power compared with the MAX test is rather small, and the MAX test has the additional advantage that it allows us to make some inference regarding which of the underlying genetic models actually holds (see [59] for further details on this point).

However, we performed our simulation for the case where both alleles occurred with the same frequency. This is frequently more or less the case in experimental populations (like, for example, in the intercross design described in Sect. 2.2.2), whereas in human genetics one has the situation that for most SNPs one allele is much more frequent than the other. For small MAFs, it is almost impossible to distinguish between additive and dominance effects (which is true for both of the definitions of dominance discussed previously). This is, perhaps, the main reason why in GWAS it is quite common to test only for additive effects.

Another difficulty is connected with the extension of the MAX test to more complex models. In Sect. 5.4, we will discuss the advantages of multi-marker models in the case of complex traits, and it seems to be rather difficult to extend the approach of the MAX test to incorporate several markers. If one is interested in assessing dominance effects, then it is much easier to incorporate the ANOVA approach into a multi-marker setting.

5.3.3 Covariates and Population Stratification

In many applications, it is the case that, apart from genetic factors, one would like to consider additional variables which have an influence on the trait of interest. Typical examples of such factors in the human population are gender and age. To model a quantitative trait Y, the previous section introduced statistical procedures in the form of a GLM with dummy variables coding the SNP genotype as regressors. It is then straight forward to extend this GLM and include the covariates in question as additional regressors. For example, a model based on an additive effect from one SNP with age and sex as covariates can be written as

$$Y_i = \beta_0 + \beta_1 G_i + \beta_2 \, \text{Age}_i + \beta_3 \, \text{Sex}_i.$$

Here Sex is a dummy variable that takes the values 1 and 0 for males and females, respectively.

The single marker tests for case-control studies presented in Sect. 5.3.1 are based on contingency tables, and it is not immediately obvious how to include covariates. One relatively simple approach is to make use of logistic regression, which was introduced in Sect. 3.3.1 and is discussed in more detail in Sect. 8.5. Consider, once again, an additive effect with age and sex as covariates, but this time we model the risk π_i that individual i is a case. The corresponding logistic regression model is

$$\log\left(\frac{\pi_i}{1-\pi_i}\right) = \beta_0 + \beta_1 G_i + \beta_2 \, \text{Age}_i + \beta_3 \, \text{Sex}_i,$$

where the left-hand side is the logarithm of the odds that individual i is a case. Estimating the coefficients in this model is carried out routinely using the maximum likelihood method (see Sect. 3.3.1). The formal analogy that exists between linear regression and logistic regression is due to the fact that both are generalized linear models (gLMs, see Sect. 8.5). This helps to unify the discussion on quantitative traits and case-control studies. If there are no covariates, then a score test for β_1 in the logistic regression model $\log\left(\frac{\pi_i}{1-\pi_i}\right) = \beta_0 + \beta_1 G_i$ is equivalent to the CAT. In this sense, logistic regression can also be seen as a generalization of the single marker tests discussed previously.

The question of whether the study population shows any kind of substructure is of particular importance in GWAS. If the population structure is not accounted for, it might give rise to so-called spurious associations [96]. Assume that a GWAS with a case-control design is performed to detect the genetic association of SNPs with some disease. Furthermore, assume that this disease has a higher prevalence in a particular subgroup of the study population than in the rest of the population. As a result, the disease will be associated with any marker whose MAF within that subpopulation differs from its MAF in the rest of the population. The association between such a marker and a disease is spurious, because, most likely, it has nothing to do with the disease itself, but is related purely to subgroup membership.

Stratification occurs whenever a population is not panmictic, i.e., whenever the assumption of random mating is strongly violated. For example, there might be members of different ethnic backgrounds within a study population. More subtle sources of stratification could include the geographical spread of a population, or mating habits based on social or class differences. The specific case of admixture populations will be discussed in detail in Sect. 5.5.

Usually, population structure can be distinguished from family structure. If the family relationship between each pair of members of a sample is known, then the conceptual framework for the data analysis is normally different and one might apply specific statistical methods for family-based studies, such as the TDT for family trios briefly discussed in Sect. 5.3.3. However, if there is kinship between some participants of a study which is unknown to the researchers, then one speaks about cryptic relatedness, a phenomenon which also has to be taken into account in GWAS analysis.

The simplest situation occurs, of course, when both the population structure and family structure are well known. As an example, consider the population from phase II of the HapMap project, which was introduced in Sect. 2.2.3. The 45 individuals from Tokyo, Japan (JPT) and the 44 individuals from Beijing, China (CHB) are unrelated. On the other hand, the group from the Yoruba tribe from Ibadan, Nigeria (YRI), and the group from Utah of European ancestry (CEU) both consisted of 30 family trios. Taking only the 60 parents from the YRI and CEU groups yields an outbred population with 4 subgroups. Thus, in the context of gLM, one could add three dummy variables (e.g. for CEU, JPT and CHB, if one wants to consider YRI as the reference group) to account for population structure. How to analyze the family trios will be discussed in Sect. 5.3.3.

Usually, the population structure is not known in advance, and one has to find a way to infer stratification from the data. A variety of methods to accomplish this task can be found in the genetics literature. Perhaps the simplest and most common approach is **genomic control** [37], which was applied in genetic association studies long before the advent of GWAS. In the case of an unstructured population and no association, the CAT statistic (5.1) is χ_1^2-distributed, but tends to be inflated when the population is stratified. The theoretical considerations discussed in [37], as well as simulation results from [105], indicate that, under relatively general assumptions on the population structure and a lack of association with the trait of interest, the CAT statistic, z_C^2, is inflated by a factor λ, such that $z_C^2 \sim \lambda \chi_1^2$. The dispersion parameter λ depends on the particular structure of the study population and equals one in the absence of stratification. As suggested by [37], genomic control uses a number of unlinked markers to estimate λ. In GWAS, one has an abundance of markers and can compute, for example, the median over the CAT statistics from a large number of randomly chosen SNPs and divide this by 0.456, the theoretical median of χ_1^2. One disadvantage of genomic control is that it only works for single marker tests and cannot be easily extended to multi-marker models.

Another popular approach to correcting for stratification is based on **principal components analysis** (PCA), a standard tool of multivariate statistics which is introduced in more detail in Sect. 10.1. More than 30 years ago, PCA was already being applied to study the geographic distribution of gene frequencies in Europe [84]. In the context of GWAS, [98] introduced the EIGENSTRAT method to account for population structure. The main idea is to perform PCA for the genotype data of all the individuals, where at least some of the leading principal components are believed to represent population stratification. EIGENSTRAT adjusts the genotypes of individuals according to these principal components and then performs chi-square tests using these adjusted genotypes. Equivalently, one can work with a gLM (logistic regression for case-control studies) and add the leading principal components as covariates. This is of particular interest in view of Sect. 5.4, because this makes it possible to incorporate adjustment for population structure into multi-marker models.

There exist a number of other ways to deal with population structure, which have recently been reviewed in [101]. One approach is based on inferring genetic ancestry using programs like STRUCTURE [97] and ADMIXTURE [3], which we will discuss in more detail in Sect. 5.5. A comparison of genomic control and inferring genetic ancestry was given in [117], where, additionally, logistic regression based on two different strategies for model selection was considered. Interestingly, logistic regression performed quite well in a simulation study, even without specifically taking population stratification into account.

A completely different strategy for avoiding problems due to population stratification was suggested by [118]. The so-called transmission disequilibrium test (TDT) is based on data from family trios, which consist of two parents together with an affected child. The principle of TDT is that the transmission of causal genetic variants is assessed by testing whether marker alleles from heterozygous parents are overtransmitted to the affected offspring, i.e., more often than expected under the hypothesis of no linkage. The TDT can be extended to situations where there is more than one offspring and thus can be seen as a special case of a pedigree-based association test. Testing for association based on family trios is less vulnerable to the influence of population structure than the population-based tests discussed above. On the other hand, it is more difficult to obtain large sample sizes under such a study design. Furthermore, it appears to be difficult to extend the TDT to include the combined effect of several markers. This is the main reason that we do not pursue tests for genetic association using family-based studies any further in this book, although it is a very important approach. For a brief review on family-based genome-wide association studies, we refer the reader to [13].

The most recent approach to addressing population structure in GWAS is based on mixed models(see Sect. 8.6). This approach was suggested a number of years ago in [129], where mixed models are not only used to consider population structure, but also to take into account closer family relationships. We will come back to such an approach in Sect. 5.4.6.

5.3.4 Multiple Testing Correction

It is still commonly accepted in the genetic research community that GWAS analysis is performed using single marker tests combined with some correction for multiple testing. All the methods discussed in Sect. 3.2 are potentially useful for this purpose, but there are certain procedures which are more popular in practice. While theoretically not the best choice, the Bonferroni procedure is widely applied, mainly because of its simplicity. Although FDR control procedures like the BH procedure have become almost standard in the area of RNA microarray analysis, they do not seem to play such an important role in GWAS analysis. One reason for this could be that, so far, the number of SNPs which have been found to be associated with a condition by GWAS is usually rather small, in which case there is no big difference between applying the Bonferroni correction or applying BH.

Also, in many projects GWAS are performed as a first stage, where they are used to select a number of interesting SNPs. These are then verified by a second stage of the study, where only the selected SNPs are genotyped in a different study population. In the case of such designs, there is a question of whether it is really necessary to control FWER so strictly in the first stage, or whether it might be preferable to allow a larger number of potential false positives to enter the second stage. The general question of how to divide resources between the first stage (GWAS) and the second stage (confirmatory study) has been addressed by [131], who provide guidelines regarding which strategies give the optimal power while simultaneously controlling the false discovery rate.

5.3.5   Rare SNPs

So far, we have only considered the association of traits with common SNPs, which are usually defined to have a MAF greater than 0.05 (or greater than 0.01 in some cases). The rationale behind looking at common SNPs is the common disease–common variant (CDCV) hypothesis mentioned previously, which assumes that common diseases are caused by a moderate number of common variants. While GWAS have been successful in detecting a number of important genes for common diseases, there has been a certain amount of disappointment in the fact that the amount of genetic heritability explained was less than expected. A number of sources which might explain this "missing heritability" have been listed in [76], among them rare variants.

There has been quite a heated debate between proponents of the CDCV hypothesis and proponents of the common disease—rare variant (CDRV) hypothesis. The latter group assume that complex diseases are caused by multiple rare variants, where the effect sizes of the rare causal variants are expected to be greater than those of common variants under the CDCV hypothesis. Meanwhile, examples of conditions conforming to each hypothesis have been described, and there seems be an understanding that both mechanisms might be of importance in explaining complex traits, depending on the genetic region involved [60]. However, only the recent advances in next generation sequencing technology have made it actually possible to assay rare variants on a genome-wide scale. Hence, the question of how to statistically analyze rare variants is becoming more and more important. Two recent reviews on statistical methods adapted to rare variants are [5, 9]. Here, we will only focus on approaches which can quite easily be interpreted in the context of model selection discussed later in Sect. 5.4.

The main difficulty in analyzing rare variants is that the simple single marker tests described in Sects. 5.3.1 and 5.3.2 will have very low power. This is due to the fact that even in a large sample only very few individuals are expected to carry rare alleles. Alternatively, one might consider multivariate models including a number of rare variants, but this approach also suffers from a lack of power, as we will see below. To date, the most popular approaches to detecting rare causal variants are based on so-called collapsing strategies, first introduced in [70, 88]. The basic idea is to combine the information from several rare variants within a given genomic region, for example, within a gene.

To formalize these ideas, consider a group of K SNPs with genotypes X_i, $i = 1, \ldots, K$, where for each SNP the minor allele is rare. There are two basic collapsing strategies which have been considered in the literature. First, the cohort allelic sums test (CAST) was proposed in [88]. The basic idea can be illustrated in a regression setting by considering the model

$$Y_i = \beta_0 + \beta_S \sum_{i=1}^{K} X_i. \tag{5.3}$$

Note that only one coefficient β_S is estimated for the combination of the K variants, whereas in a more standard multivariate model one would have K coefficients $\beta_1, \ldots, \beta_K$, one coefficient for each variant. The LRT test statistic for the full multivariate model is approximately χ^2 distributed with K degrees of freedom, whereas the LRT test statistic for β_S has only one degree of freedom. One would therefore hope that a test based on model (5.3) is more powerful in detecting a region where many rare variants have an effect on the trait than the full multivariate model.

The second approach might be referred to as collapsing per se, where one uses an indicator variable which is defined to be equal to 1 if and only if at least one of the genotypes includes a rare allele, i.e.,

$$Y_i = \beta_0 + \beta_C \left(1 - \prod_{i=1}^{K} \mathbf{1}_{X_i=-1} \right). \tag{5.4}$$

Again, only one parameter has to be estimated for the group of K SNPs, and the LRT for β_C has only one degree of freedom. Compared to performing K single marker tests, one now only has to perform one test, reducing the burden of statistical testing. Furthermore, one ends up with an enriched signal, because even when each variant itself is rare, the combined information of K markers will result in a signal which will be observed with larger frequency.

The combined multivariate and collapsing (CMC) method suggested by [70] makes use of the second collapsing strategy (5.4). However, CMC has the additional feature of generating groups of variants not only based on genomic region, but also by clustering SNPs with a similar MAF. Therefore, for any given genomic region, several different groups of SNPs are considered for which information is collapsed separately, and then a multivariate test is performed for these various collapsed variables.

Both of these approaches implicitly assume that all of the combined rare variants have the same kind of effect on the trait. For example, in a case-control design, it is assumed that the minor alleles of all SNPs either increase the risk of the disease, or that all the minor alleles are protective. In the setting of regression, this corresponds to the situation where all the β_i in the multivariate model have the same sign. However, in practice, this is not necessarily true and it might be the case that some rare variants correspond to risk alleles, whereas others are protective. In such a situation, collapsing will inevitably yield a loss of power. One possible strategy to overcome this problem was developed in [53], where the coding of the genotype X_i is changed when the sign of β_i in a multivariate model is negative. After recoding the SNPs, which more or less guarantees that all the effects are aligned in the same direction, a sum test is performed based on the model (5.3). To protect against inflated type I error rates due to the data driven recoding of SNPs, p-values are then obtained using permutation tests (see Sect. 3.2.2).

5.4   Model Selection

In this section, we are going to discuss in detail how to perform the analysis of GWAS using model selection. The first subsection will discuss in depth what can be gained from multilocus models compared to single marker analysis. The three subsequent subsections introduce different approaches to model selection for GWAS that are implemented in software packages which are rather easy to use for real GWAS analysis, namely MOSGWA [39], GWASELECT [55], and HYPERLASSO [58]. In Sect. 5.4.5, we compare the performance of these three strategies of model selection and discuss some other approaches suggested in the literature.

5.4.1   Motivation

Single marker analysis, as discussed in Sect. 5.3, still appears to be the standard form of analysis for GWAS data. To a certain extent, this is due to the computational and conceptual challenges that more sophisticated methods of statistical analysis impose in the context of such high-dimensional data. However, GWAS are performed with the intention of gaining an understanding of the genetic factors associated with complex traits. By definition, complex traits involve a larger number of genetic risk factors, and we will demonstrate that in this situation single marker tests have some serious drawbacks. To this end, we will base our discussion on results from [48], which deal with GWAS for quantitative traits. Case-control studies will be discussed afterwards.

We assume that for n different individuals the genotype X_{ij} of m SNPs is known, where $X_{ij} = \pm 1$ for the two possible homozygote states, and $X_{ij} = 0$ in the case of heterozygosity. In our notation, $i \in \{1, \ldots, n\}$ is the index for individuals and $j \in \{1, \ldots, m\}$ is the index for SNPs. For the purpose of illustrating the basic concepts, the individuals are assumed to come from a perfectly outbred population, which is homogeneous in the sense that there is no underlying population structure. Some quantitative trait Y_i is influenced by k causal variants. For the sake of simplicity, we consider the following additive model (compare with Eq. (3.18)):

$$Y_i = \beta_0 + \sum_{j \in S^*} \beta_j X_{ij} + \epsilon_i , \quad \epsilon_i \sim \mathcal{N}(0, \sigma^2) \text{ i.i.d.}, \tag{5.5}$$

where S^* is the set of k causal SNPs.

In reality, several aspects of this model are oversimplified. First of all, one does not expect that the influential genetic factors are actually among the SNPs one has genotyped. It is much more likely that a genotyped SNP is found within the neighborhood of some causal genetic factor and, due to linkage disequilibrium, the genotype of this SNP is strongly correlated to the genetic factor. The linear model thus reflects the association of such a SNP with the trait of interest due to this correlation with the genetic effect. A second simplification is that the model only includes additive effects. As we have seen in Sect. 5.3.2, it might be the case that some dominance effects play a role. Furthermore, epistatic effects could be important, which are usually modeled as interaction effects between SNPs. Note that it would be conceptually simple to include dominance and interaction effects in the model, as we have already seen in the context of QTL mapping in Chap. 4. Finally, the assumption that the error terms are i.i.d. and normally distributed appears to be a convenient choice, which might not always hold. However, as explained in Sect. 8.4, GLMs have a certain robustness with respect to small violations of the normality assumptions.

While one has to be aware of the limitations of the simple linear additive model, it has the advantage that its statistical properties are extremely well understood and the analysis of many of its features is rather easy. Thus it can serve the purpose of shedding light on certain intrinsic difficulties that occur when applying single marker tests to detecting associations in the case of complex traits. In particular, we will see that single marker tests tend to have unnecessarily low power, they have major problems with appropriately ranking causal SNPs, and finally they are prone to detecting false positive SNPs.

The first problem, low power, is quite easy to explain. When performing a single marker test for a given SNP, none of the effects of the other causal factors on Y are considered and this inflates the residual variance. This, in turn, leads to a substantial decrease in the values of the test statistic. The resulting loss of power can be very large and increases with the number of causal variants and the magnitude of their effects.

To illustrate the low power of single marker analysis, we consider an extremely simple example where the regressors of the model (5.5) are assumed to be orthogonal. It is then particularly easy to compute the power of marginal F-tests which test the hypothesis $H_0 : \beta_j = 0$ against $H_A : \beta_j \neq 0$ (see Sect. 8.3.5 for the specification of the F-test, and [48] for the calculations of the exact power). In Fig. 5.6, one can see to which extent power is lost when marginal F-tests are applied to models of increasing complexity.

Fig. 5.6 Theoretical power for marginal F-tests depending on the effect size. The correct model is applied only when $k = 1$. Otherwise, one observes a severe loss of power as the number of regressors k in the true model increases

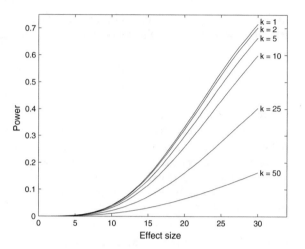

The plot also gives an idea about the loss of power which should be expected when single marker tests are applied to detect SNPs which are associated with complex traits. The greater the number of independent genetic factors that have an influence on the trait, the more inappropriate it will be to try to find them using single marker tests. This observation seems to have direct implication on the widely discussed phenomenon of missing heritability in GWAS, which refers to the fact that the SNPs which have so far been detected by GWAS explain only a very small proportion of the variability which one would expect to be explicable by genetic factors. Many explanations have been given for this, including the contribution of rare variants, epistasis, epigenetics, etc. [41, 76]. However, recently it was reported in [128] that the missing heritability of human height can be explained by the joint influence of many common variants, which are not detected by the statistical methods commonly in use at present. This is one strong motivation for the use of more sophisticated statistical methods, such as multiple regression combined with model selection.

The second problem mentioned above is that single marker tests tend to have substantial problems with ranking causal SNPs correctly. This is perhaps even more undesirable than having low power, given the fact that it is common practice in GWAS to report the top ranking SNPs. This problem is illustrated in Fig. 5.7a, which presents the results from a simulation study based on real SNP data based on 649 individuals from the POPRES sample. From a total of 30,9788 SNPs, a subset of 40 was chosen to be causal. All of the causal SNPs are common with minor allele frequency MAF > 0.3. They are distant from each other and the pairwise correlations range between -0.12 and 0.1, which corresponds well to the expected range of correlations for a random sample of SNPs which are in complete linkage equilibrium. Further details of this study are given in [48].

In Fig. 5.7a, the power to detect causal SNPs with single marker tests is compared to a more advanced statistical approach based on strategies for model selection. Single marker tests are corrected for multiple testing using the standard

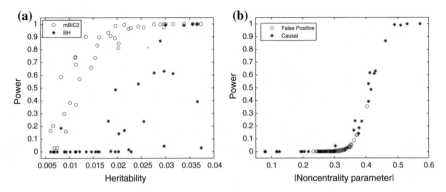

Fig. 5.7 a Dependence of the detection rate on the individual SNP heritability for single marker tests (BH) and for the model selection procedure based on mBIC2. **b** Detection rate for causal SNPs and false positives under single marker tests (BH) as a function of $|v|$, the absolute value of the noncentrality parameter of the marginal F-test

Benjamini–Hochberg procedure (BH from Sect. 3.2.3) aimed at controlling the False Dicovery Rate (FDR) at the standard level of 0.05. The more sophisticated approach is based on multiple regression models using the mBIC2 selection criterion from Sect. 3.3.4, which is also based on the principle of controlling FDR.

In our study, the observed FDR for the single marker test procedure was on average 0.16, while the FDR of the mBIC2 procedure was at the substantially lower level of 0.04. In spite of this, the approach of model selection was much more powerful than single marker tests, as illustrated in Fig. 5.7a. This is a direct consequence of the loss of power of single marker tests discussed previously. Here, the power of detecting a causal variant is plotted as a function of "individual heritability," defined as $h_j^2 := \beta_j \text{Var}(X_j)/\text{Var}(Y)$. Individual heritability is simply the proportion of variation which is explained by a single causal SNP and can be directly interpreted as a scaled effect size, which takes into account the variation in the SNP genotype. The power of mBIC2 behaves as expected, in the sense that it gets consistently larger as heritability increases. On the other hand, the behavior of the power of the BH procedure as a function of heritability is quite erratic. For example, the causal SNP with the largest individual heritability is detected by BH in only 3 of 1000 simulations. On the other hand, several SNPs with considerably smaller heritability are detected much more often.

The explanation of why single marker tests have problems with ranking causal SNPs is slightly more subtle than the explanation of the loss of power discussed above. When trait Y is influenced by the k causal variants $j_1, \ldots, j_k \in S^*$ according to model (5.5), then the sample covariance between the trait and the genotype of the first causal SNP is given by

$$\text{Cov}(Y, X_{j_1}) = \beta_{j_1} \text{Var}(X_{j_1}) + \sum_{r \in S^*, r \neq j_1} \beta_r \text{Cov}(X_{j_1}, X_r) + \text{Cov}(X_{j_1}, \varepsilon). \quad (5.6)$$

Even if the first SNP is in complete linkage equilibrium with all the other causal SNPs, the sample covariances between the respective genotypes will almost never be equal to zero. While individual sample covariances will typically be very small, the net effect due to the sum of $k - 1$ such components can overshadow the first element in the sum, i.e., the effect of the SNP itself.

Figure 5.7b illustrates this situation by plotting the power of detecting causal SNPs against the noncentrality term defined below, which is crucial in determining the observed detection rate. In [48], the distribution of the marginal F-test statistic is studied more closely, and it is shown that if model (5.5) is correct, then the sum of squares MSS corresponding to the test $H_0 : \beta_{j_1} = 0$ has a noncentral χ^2 distribution with noncentrality parameter

$$\nu_{j_1} := \frac{\sum_{r \in S^*} \beta_r \mathrm{Cov}(X_{j_1}, X_r)}{\sigma \sqrt{\mathrm{Var}(X_{j_1})}}. \tag{5.7}$$

It is worth comparing the noncentrality parameter with the covariance term (5.6). Note that ν_{j_1} is directly proportional to $\mathrm{Cov}(Y, X_{j_1}) - \mathrm{Cov}(X_{j_1}, \varepsilon)$, which is simply the nonrandom contribution to the covariance between X_{j_1} and Y. The sigmoidal curve in Fig. 5.7b clearly indicates that this parameter is responsible for the detection rates of BH, and not the individual heritabilities, which one would actually be interested in. In our simulation study, the pairwise correlations between causal SNPs was deliberately chosen to be rather small, so that all the causal SNPs could be considered as being independent in a statistical sense. Still, for some SNPs the weighted sum of the correlations in (5.7) is large enough to dominate the absolute value of the noncentrality parameter. So what we observe is not really a problem of multicollinearity, but rather the effect of sample correlations between causal markers with small absolute values, which only start to play a role when the number of causal markers becomes relatively large. It is particularly disconcerting to note that the pattern of these sample correlations in a particular study will most likely be quite different from the pattern in another study, which provides one explanation of why the results of GWAS have very often not been replicable in follow-up studies.

The third remarkable shortcoming of single marker tests in the case of complex traits is related to the detection of false positive SNPs, which are not at all correlated with any causal SNPs. The explanation of this phenomenon is the same as above, i.e., for some SNPs unassociated with the trait, it just happens by chance that the weighted sum of correlations in expression (5.7) is rather large. Figure 5.7b also illustrates the detection rate for all the SNPs which were detected at least once as false positives in the simulation study as a function of the magnitude of the noncentrality parameter. Detected SNPs were classified as a false positive, when the correlation with any of the causal SNPs did not exceed 0.7. In Fig. 5.7b, we can observe six false positives with a detection rate above 5 %. Only one of them is relatively strongly correlated to a causal SNP ($|r| = 0.57$), the other five are more or less uncorrelated to any of the 40 causal SNPs. The most commonly detected, by far, false positive has a detection rate of more than 35 %, but its maximal correlation with any of the 40 causal SNPs is 0.26. The second most frequently detected false positive has a detection rate of 13 %, but its maximal correlation with any of the 40 causal SNPs is only 0.11, which suggests that it is in linkage equilibrium with all causal markers.

The reason for the frequent detection of these false positives is that in our sample they turn out to be correlated to some linear combination of causal variants. This situation is not exceptional, but bound to occur quite frequently in GWAS. Due to the large number of markers, almost every vector containing the genotypes of a specific SNP can be expressed as a linear combination of vectors corresponding to other, usually unrelated, SNPs. In some cases, such a linear combination will include causal SNPs with the same sign, leading to a false discovery. This explains why the Benjamini–Hochberg procedure fails to control FDR at the nominal level of 0.05 in our simulation study.

In summary, we have seen that there are several reasons which speak strongly against the use of single marker tests to analyze GWAS data. In the following sections we will introduce various approaches to model selection, where we focus on those for which reliable software is available. So far, we have only discussed quantitative traits, which can be modeled using a simple linear regression model (5.5). The majority of GWAS are concerned with case-control (or case random) designs, where the trait can take only two different values (affected vs. non-affected). This kind of study is routinely modeled via logistic regression, a special case of a generalized linear model [81]. A more detailed discussion is given in Sect. 8.5 of the Appendix.

For a case-control design, let $\pi_i = P(Y_i = 1)$ denote the probability that the ith individual is affected. Then according to the logistic regression model (3.21), the logarithm of the odds is assumed to be a linear function of causal factors, as follows:

$$\log\left(\frac{\pi_i}{1 - \pi_i}\right) = \beta_0 + \sum_{j \in S^*} \beta_j X_{ij} =: [X\beta]_i. \tag{5.8}$$

Given data from n individuals, the likelihood of a specific model is given by

$$\mathcal{L}(\beta \mid Y, X) = \prod_{i=1}^{n} \pi_i^{Y_i} (1 - \pi_i)^{1 - Y_i} = \prod_{i=1}^{n} \frac{\exp\left(Y_i [X\beta]_i\right)}{1 + \exp\left([X\beta]_i\right)},$$

and thus the whole machinery of likelihood based or Bayesian model selection introduced in Sect. 3.3 can be applied.

5.4.2 HYPERLASSO

HLASSO (short for HYPERLASSO), which was introduced in 2008 [58], was among the first software packages which were actually able to analyze GWAS data using model selection. HLASSO essentially takes a Bayesian approach to model selection using shrinkage priors. This avoids time-consuming MCMC methods for computing the posterior probabilities of models, but instead is based on a very interesting optimization scheme to maximize the mode of the posterior distribution.

The basic idea of [58] is as follows: Consider a saturated model, including all the available SNPs, with parameter vector $\beta = (\beta_0, \beta_1, \ldots, \beta_m)$, together with some prior distribution with joint density $f(\beta)$. According to Bayes formula, the posterior density of β is then computed as

$$f(\beta \mid Y, X) = \frac{\mathcal{L}(\beta \mid Y, X) f(\beta)}{\int_{\beta'} \mathcal{L}(\beta' \mid Y, X) f(\beta') \, d\beta'}. \tag{5.9}$$

The denominator on the right-hand side of Eq. (5.9) is simply a constant and can be neglected when trying to find the mode of the posterior density. Taking logarithms and introducing the notation $p(\beta) := -\log(f(\beta))$ yields

$$\log f(\beta \mid Y, X) = \log \mathcal{L}(\beta \mid Y, X) - p(\beta) + const. \tag{5.10}$$

Thus the problem of finding the posterior mode has been reformulated in terms of maximizing a penalized loglikelihood. Compare the expressions for AIC (3.25) and BIC (3.28) with Eq. (5.10). Multiplying the first two expressions by $-1/2$ transforms these selection criteria into penalized loglikelihoods which have to be maximized, and which are of the same general form as Eq. (5.10).

For AIC and BIC, the penalty $p(\beta)$ depends on $\|\beta\|_0$, the number of coefficients which are nonzero. The exact form of the penalty function $p(\beta)$ in Eq. (5.10) obviously depends on the choice of the prior distribution. Shrinkage priors have the property that they have cusps (sharp peaks) at the origin. As a result, many components of the β vector yielding the posterior mode tend to be equal to 0, and this forms a basis for model selection (see [92, 122]).

HLASSO has been implemented with two different families of shrinkage priors, where it is always assumed that the prior distributions for different coefficients β_j are independent, i.e., $f(\beta) = \prod_j f(\beta_j)$. The first type of prior considered in [58] is the double exponential distribution (DE, see also 6.4), a one-parameter family of shrinkage priors, which is quite common in Bayesian statistics and has density function

$$f_{DE}(\beta_j \mid \xi) = \frac{\xi}{2} \exp\left(-\xi |\beta_j|\right) = \int_{\sigma^2=0}^{\infty} \sigma^{-1} \phi(\beta_j/\sigma) Ga(\sigma^2 \mid 1, \xi^2/2) d\sigma^2. \tag{5.11}$$

Figure 5.8 shows the density function and the logarithm of the density function of the DE for various choices of ξ. Note that the log densities are just piecewise linear functions. Plugging this log density into (5.10) thus yields

$$\log f(\beta \mid Y, X) = \log \mathcal{L}(\beta \mid Y, X) - \xi \sum_{j=1}^{m} |\beta_j| + const,$$

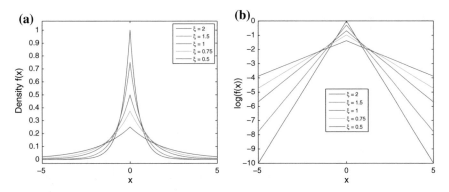

Fig. 5.8 **a** Density of the double exponential distribution for different values of the parameter ξ. **b** Logarithm of the density

and we immediately conclude that maximizing this expression is exactly the same problem as minimizing expression (3.37). Therefore, deriving the posterior mode for a double exponential prior is equivalent to the lasso procedure as introduced in Sect. 3.3.5, where the scaling parameter ξ takes the role of the tuning parameter λ from lasso. This connection with a Bayesian approach to model selection had already been pointed out in the original article on lasso [122].

The second class of priors discussed in [58] is the normal exponential gamma (NEG) distribution, which can be written as a scale mixture of normal distributions, as follows:

$$f_{NEG}(\beta_j|\eta,\gamma) = \int_{Psi=0}^{\infty} \int_{\sigma^2=0}^{\infty} \sigma^{-1}\phi(\beta_j/\sigma)Ga(\sigma^2|1,\Psi)Ga(\Psi|\eta,\gamma^2)d\sigma^2 d\Psi$$
$$\propto \exp\left(\beta_j^2\gamma^{-2}/4\right) D_{-(2\eta+1)}(|\beta_j|/\gamma). \tag{5.12}$$

The NEG is slightly less commonly applied than the DE, but also serves as a shrinkage prior. In fact, the peak of the NEG density curve at $\beta_j = 0$ is much sharper than DE's peak, which is the reason behind the name HYPERLASSO. The exact form of f_{NEG} depends on the shape parameter η and the scale parameter γ. If η and γ both increase in such a way that $\xi = \sqrt{2\eta}/\gamma$ remains constant, then the NEG converges toward the DE. The second way of mathematically describing f_{NEG} involves the parabolic cylinder function D_a with parameter a, and a constant of integration, which we only indicate by the proportional relationship in (5.12). Fast algorithms are available to evaluate parabolic cylinder functions, see [50].

One outstanding feature of HLASSO is the fast algorithm to optimize expression (5.10) in combination with an NEG prior, which is based on the CLG algorithm [12]. This allows HLASSO to perform model selection over the whole set of SNPs from GWAS without preselection based on marginal tests. Another interesting feature is that in [58] some arguments are given regarding how to choose the parameters of the prior such that the type I error rate is controlled. We will see in Sect. 5.4.5 how different parameter choices affect the FWER of HLASSO.

5.4.3 GWASelect

A second software package which is capable of performing model selection for GWAS is GWASelect [55], which combines several different ideas for model selection in high dimensions. In the first step, GWASelect applies the idea of sure independence screening (SIS) [43], which means that it performs marginal tests (Cochran–Armitage) and considers only those markers which have the smallest marginal p-values for further analysis. According to SIS theory, the number of SNPs is thus reduced from m down to $\lfloor 0.9n/4 \log n \rfloor$, where n is the number of individuals. Model selection among logistic regression models based on the remaining SNPs is then performed using LASSO, resulting in a model $\mathcal{M}_1^*$, where the tuning parameter λ from (3.37) is determined via cross validation.

Inspired by the idea of iterated SIS (ISIS) from [43], GWASelect then performs a second round of tests to select markers. The marginal Cochran–Armitage trend tests of the first SIS are now substituted by score tests based on the model $\mathcal{M}_1^*$. In particular, such score tests are performed for all the SNPs which are not included in model $\mathcal{M}_1^*$, and the $\lfloor 0.05n/4 \log n \rfloor$ top ranking SNPs are combined with the SNPs from model $\mathcal{M}_1^*$ to again perform LASSO based model selection resulting in model $\mathcal{M}_2^*$. The second step is repeated one more time, with score tests based on $\mathcal{M}_2^*$ instead of $\mathcal{M}_1^*$.

The idea behind this iterative procedure is that the first SIS can only detect SNPs with strong marginal effects, whereas the second and third round of ISIS have the potential to identify SNPs which are marginally uncorrelated to the trait of interest, but still might be significant conditional on the effects of the SNPs presently in the model. Score tests have the advantage that they are computationally much less intensive than other tests. Likelihood ratio tests require that an optimization problem is solved for each marker to be tested, while score tests only require the ML estimate of the model on which one would like to condition (see Sect. 3.3.1). The computation of the conditional score test statistic is then rather cheap (see [55] for details).

Another ingredient of GWASelect is the stability selection strategy suggested by [83]. The ISIS procedure described above is performed 50 times on different subsamples, where each subsample is obtained by simply drawing 50 % of the cases and 50 % of the controls at random. This results in 50 different models, and the final model presented by GWASelect includes only those markers which were selected with sufficiently large relative frequency among these 50 models. According to [55], the specific choice of this threshold, ξ, on the relative frequency has no dramatic effect on the final model. However, we will see in Sect. 5.4.5 that this does not seem to be the case according to our own simulations.

5.4.4 MOSGWA

The MOSGWA software package, described in [39] and available at http://mosgwa. sourceforge.net/, uses a heuristic approach to choose a near-optimal linear or logistic regression model, as appropriate, according to the mBIC2 criterion (see [48] for a background to using adaptations of the Bayesian Information Criterion to model selection for GWAS). Suppose there are m SNPs, n individuals and we have a logistic regression model M for the presence or absence of a certain condition based on k_M SNPs. The criterion used to select a model is to minimize mBIC2, where

$$mBIC2 = -2 \log L_M^* + k_M \log(nm^2/4) - 2 \log(k_M!).$$

Here L_M^* is the Firth corrected likelihood of the data under model M and $L_M^* = L_M \sqrt{|I(\theta)|}$, where $|I(\theta)|$ is the determinant of the Fisher information matrix for the variables included within the model. It should be noted that this correction of the likelihood function is used to avoid the phenomenon of separation. Separation can occur when the number of regressors is large compared to the number of individuals and certain combinations of regressors split the sample into two groups, one that has the condition in question and one that does not. In such a case, the estimate of a parameter becomes excessively large. The use of this corrected likelihood avoids this problem. It should be noted that this problem of parameter inflation is not present under the GWASelect or HYPERLASSO methods, since in those cases the likelihood function is penalized according to the values of the parameters and not the number of nonzero parameters. In the case of linear regression, MOSGWA uses the standard mBIC2 criterion.

Denote the model obtained at the end of the ith stage of the algorithm by M_i. The ith stage of the fast stepwise search (FSS) algorithm can be described as $M_i = FSS(M_{i-1}, Test, Criterion)$, where M_0 is the null model (i.e., M_0 assumes that no SNP is associated with the disease). Hence, the model obtained at the ith stage depends on the model at the beginning of the stage, a test which is used to rank the importance of the SNPs and the criterion used to choose the model. Each stage is carried out as follows:

(i) **Ranking**: The significance of each SNP is ranked according to a given statistical test. The SNPs to be considered are categorized into two groups. Group G_1 consists of the m_1 most highly ranked SNPs and G_2 consists of the m_2 most highly ranked SNPs, where $m_2 > m_1$. By default $m_1 = 350$ and $m_2 = 5000$, although the specific choice of these values has very little influence on the results.

(ii) **Forward Step**: Each SNP in G_1 is considered in turn starting from the highest ranked SNP. A SNP is added to the model, if and only if the resulting model is an improvement on the present model according to the criterion used.

(iii) **Exchange Step**: We then check whether a SNP in the model can be replaced by one of the 49 nearest neighboring SNPs on either side from the set G_2, starting with the highest ranked SNP in the model. If an improvement can be made by switching a SNP, then the best swap, according to the criterion, is made.

(iv) **Backward Step**: If removal of a SNP leads to an improvement according to the criterion used, then the removal leading to the greatest improvement is implemented. When no improvement is made, the SNP which explains the least variance is removed and another backward step is attempted. If three such steps do not lead to an improvement, then the best model found so far is used as the starting point in the next stage. Otherwise, the best improvement found is accepted as the new model and removal continues.

The forward and backward steps used in the algorithm are very similar to those used in standard stepwise regression procedures. The exchange step takes into account that any association between the condition of interest and a SNP in the model may result from the association between two neighboring SNPs and not necessarily from a direct association between the condition and the SNP in the model.

When the appropriate model is logistic regression, the default MOSGWA algorithm uses three stages, defined as follows:

$$M_1 = FSS(M_0, Cochran - Armitage, mBIC_{60})$$
$$M_2 = FSS(M_1, Score\ Test, mBIC_{60})$$
$$M_3 = FSS(M_2, Score\ Test, mBIC2)$$

Hence, in the first stage the ranks of the SNPs are based on the statistic for the Cochran–Armitage test. However, once a model has been built, it makes sense to measure the importance of a SNP taking into account the effect of the other SNPs in the model. Hence, in the final two steps, the score test based on the current model is used to rank the SNPs. When a linear regression model is required, then standard Student t-tests may be used to rank the SNPs in the first stage and afterward the appropriate score tests are used.

The criterion used in the first two stages is minimization of

$$mBIC_{60} = -2\log L_M^* + k_M \log(nm^2/60).$$

This function uses a smaller penalty on the dimension of the model. Hence, in the initial stages, the model chosen is likely to be overly complex. This is based on the intuition that the initial filter during the analysis of GWAS should choose too many significant variables rather than too few. In the final stage, the standard mBIC2 criterion for model selection under the assumption of sparsity is used. Hence, the final stage tends to trim the dimension of the model down to an appropriate size.

5.4.5   Comparison of Methods

As well as presenting the MOSGWA procedure, [39] also compares this procedure with the HLASSO and GWASelect procedures described above. This was done using two types of simulations, together with data from the Wellcome Trust Case Control consortium, [32]. The first set of simulations were carried out under the null hypothesis that no SNP is associated with a condition. The real SNP data used for these simulations came from the POPRES sample, [90]. This sample is composed of 4077 individuals and covers almost 150,000 SNPs on Chromosomes 1–6. In order to see how the number of SNPs considered affects the FWER, simulations were carried out for the following four problems: (i) SNP data from Chromosome 1, (ii) SNP data from Chromosomes 1 and 2, (iii) SNP data from Chromosomes 1–4, (iv) SNP data from Chromosomes 1–6. In each case, 200 simulations were carried out in which the individuals were randomly assigned to the case or the control group (each with probability 0.5).

Since the GWASelect procedure often reports SNPs which are highly correlated, one should limit the number of discoveries to the number of "independent" false discoveries. In order to do this, the SNPs were clustered into classes of linked SNPs, as described in [46], and two discoveries were defined to be independent if and only if the SNPs came from different clusters. This phenomenon was not observed for either the MOSGWA or the HLASSO procedures. He and Lin [55] suggest that the threshold ξ to be used in the GWASelect procedure should be between 0.1 and 0.2 (a SNP is reported whenever the proportion of the generated models in which it appears is at least ξ. The simulations in [39] indicate that such values did not control the FWER at all. The mean number of detections did decrease as the number of chromosomes considered increased. However, when Chromosomes 1–6 were considered, the mean number of detections was 4.19 when $\xi = 0.2$ and 32.22 when $\xi = 0.1$. Using $\xi = 0.3$ gave some control of the FWER, but it was clearly higher than the FWER obtained under either the HLASSO or the MOSGWA procedures.

The latest software for the HLASSO procedure allows the user to select a nominal significance level α, which here corresponds to the FWER. Simulations showed that the actual significance level is lower than the nominal significance level. The mean number of detections (which is an upper bound on the probability of making at least one false detection) was always lower than the nominal significance level. The HLASSO procedure with $\alpha = 0.3$ gave the most similar results to the MOSGWA procedure, which gave stable control of the FWER at close to 0.1. For this reason, in the remaining simulations and tests, the HLASSO procedure with $\alpha = 0.3$ was used.

In addition to the three procedures considered above, single marker tests were carried out which took into account the first four principal components of the SNP data, which reflect structure within the population. This analysis was carried out using the PLINK package, [102]. The FWER was controlled to be 0.05 and the simulations showed that the FWER is controlled at very close to the nominal level. Based on these simulations, the mean number of causal SNPs found under the null

Table 5.4 Mean number of falsely detected SNPs under the null hypothesis using the compared procedures

No. of SNPs	MOSGWA	HLASSO $\alpha = 0.3$	HLASSO $\alpha = 0.2$	HLASSO $\alpha = 0.1$	GWA select $\xi = 0.3$	Single marker
27,520	0.16	0.22	0.12	0.03	2.34	0.04
56,629	0.13	0.13	0.07	0.04	1.51	0.04
103,348	0.13	0.18	0.11	0.04	1.06	0.04
149,478	0.13	0.16	0.05	0.03	0.79	0.06

hypothesis using the procedures described above are presented in Table 5.4. As noted, these mean values are upper bounds on the empirical FWER. Also, the results for the GWASelect with $\xi = 0.1, 0.2$ are not given, since these procedures did not control FWER in any sensible manner.

Second, three sets of simulations were carried out based on the assumption of a complex trait. The number of causal SNPs was defined to be 6, 12 and 24, respectively. These causal SNPs were common (MAF > 0.05) and evenly distributed over the six chromosomes. No pair of causal SNPs was significantly correlated ($\rho < 0.1$ in each case). The probability of an individual having a disease was described by a logistic regression model, in which the effect sizes were in the interval $[0.2, 0.28]$, i.e., intermediate in size. The intercept in this model was chosen so that the number of cases and controls in each simulation were similar. In each case, half of the causal SNPs were removed. This mimics the situation where the SNPs observed are not causal, but may be linked to a causal SNP. It should be noted that all of the SNPs removed from the data set were in strong linkage disequilibrium with a number of other SNPs.

A detection was defined to be true if and only if the SNP detected was strongly correlated to the causal SNP, where SNPs are stated to be strongly correlated when they belong to the same C-cluster according to the clustering algorithm in [46]. The power of a procedure is defined to be the expected proportion of causal SNPs found. Estimates of the power and the FDR for the four procedures considered are given in Table 5.6. It should be noted that when there are six causal SNPs, the GWASelect procedure with a threshold of $\xi = 0.1$ has a slightly higher power than the MOSGWA procedure. However, this high power is bought at the cost of a very high FDR, 0.77. As the number of causal SNPs increased, the power of the GWASelect procedures clearly fell. Hence, in the opinion of the authors, this is not a suitable procedure for finding genetic effects of either simple or complex traits. The performance of MOSGWA and single marker tests seem very robust to the number of causal traits with MOSGWA having a slightly higher power and a comparable FDR. On the other hand, the power of HLASSO declines as the number of causal SNPs increases. This is not unsurprising, since the HLASSO procedure is aimed at the more conservative approach of controlling the FWER than controlling the FDR. The single marker tests applied the Benjamini–Hochberg procedure and thus were specifically aimed

Table 5.5 Empirical power and FDR of the compared procedures for simulations of a complex trait with effects of intermediate strength

	MOSGWA	HLASSO $\alpha = 0.3$	GWA select $\xi = 0.3$	Single marker
6 causal SNPs				
Power	0.67	0.67	0.41	0.59
FDR	0.19	0.27	0.16	0.14
12 causal SNPs				
Power	0.68	0.59	0.28	0.63
FDR	0.13	0.17	0.08	0.16
24 causal SNPs				
Power	0.65	0.52	0.24	0.62
FDR	0.13	0.09	0.09	0.16

Table 5.6 Number of significant SNPs and regions (in brackets) found by the methods compared to be associated with a set of seven conditions

Condition	WTCCC	MOSGWA	HLASSO	GWA select $\xi = 0.3$
BD	(1)	1 (1)	1 (1)	1 (1)
CAD	(1)	2 (2)	3 (2)	3 (2)
HT	(0)	1 (1)	1 (1)	1 (1)
IBD	(9)	17 (16)	12 (8)	12 (5)
RA	(3)	11 (5)	12 (2)	1 (1)
T1D	(7)	25 (11)	22 (4)	12 (2)
T2D	(3)	2 (2)	3 (2)	4 (2)

at controlling the FDR. Similarly, the MOSGWA procedure is based on the mBIC2 criterion, which under the assumption of sparsity acts very much like FDR control (Table 5.5).

Finally, these methods were used to analyze real data from the Wellcome Trust Case Control Consortium [32]. These data covered about 2000 cases for each of seven conditions: (i) bipolar disorder (BD), (ii) coronary heart disease (CAD), (iii) hypertension (HT), (iv) Crohn's disease (IBD), (v) rheumatoid arthritis (RA), (vi) type 1 diabetes (T1D), (vii) type 2 diabetes (T2D). These case groups were individually compared with a common control group, which contained about 3000 individuals. The number of significant regions for each disease detected by this Consortium using single marker tests and the number of significant SNPs and significant regions detected by the procedures compared are given in Table 5.6.

As expected from the simulations carried out above, the differences between MOSGWA and HLASSO are most visible with regard to the analysis of the most complex of these conditions (Crohn's disease, rheumatoid arthritis and type 1 diabetes). In these cases, MOSGWA confirms all the regions which were detected by the WTCCC study and indicates other regions, which had not been found to be significant using other procedures. The HLASSO procedure misses several of the regions detected by the original analysis. It should be noted that although HLASSO seems to detect a similar number of SNPs to MOSGWA, using HLASSO the SNPs detected are more often in the same region as each other. This is likely to be due to the shrinking of the regression factors inherent in the procedure. This means that a regression coefficient may underestimate the degree of association between a SNP and a condition and the addition of a neighboring SNP into the model is often required to compensate for this effect. The reason why MOSGWA detects more significant regions is probably due to the fact that, unlike MOSGWA, HLASSO is aimed more at controlling FWER rather than FDR, as discussed when considering the results of the simulations based on a complex trait. From these results, it can be seen that MOSGWA can be recommended for the analysis of GWAS, particularly when it is thought that the condition of interest is complex.

5.4.6   Mixed Models

Unlike the use of principal components analysis, briefly considered in Sect. 5.3.3, which models population structure in a somewhat abstract manner, mixed models describe population structure by estimating the matrix of pairwise relatedness between individuals on the basis of the SNP data. One major advantage of such approaches is that they can reflect subtle graduations in the degree of relatedness between individuals within a population which occur naturally, rather than splitting the population into distinct subpopulations. The Efficient Mixed-Model Association (EMMA) algorithm, [61], first estimates this relatedness matrix. It then estimates the matrix of covariances between the phenotypes, which is used to model the effect of genetic relatedness on the phenotypes. Once this effect has been taken into account, score tests may be used to detect genetic effects.

One problem with such an approach is the numerical complexity of the algorithm. In GWAS, the sample size is normally rather large and so the estimation of the relatedness between each pair of individuals involves the calculation of a very large matrix based on a huge number of SNPs. This means that using the whole data set to estimate the relatedness matrix is impractical. For this reason, a simplified version of this algorithm, the so called EMMAX (EMMA eXpedited) algorithm was introduced in [62].

Genomic control, briefly described in 5.3.3, was used to check whether the EMMAX algorithm does actually account for the effects of population structure based on ten phenotypes. In each case, after correcting for the population structure, there was no significant inflation of the test statistics. Application of PCA to account for population structure resulted in significant inflation being removed in one of these ten cases. This suggests that the EMMAX algorithm is successful in accounting for the effects of population structure. It is also argued that application of genomic control to account for the inflation of test statistics due to the population structure does not work as well as the mixed-model approach, since genomic control does not explicitly model the structure of the population and genomic control can lead to spurious deflation of test statistics and consequently to a reduction in power.

One of the simplifications used in the EMMAX algorithm is that the variance parameters are estimated under the null hypothesis, whereas EMMA estimates these parameters under the alternative. This will generally result in the shrinkage of the test statistics. However, a small scale study indicates that the resulting loss in power is small and compensated for by a vast reduction in the computation time required (the CPU time required to analyze the WTCCC data set is several hours rather than several years).

The FaST-LMM (Factored Spectrally Transformed Linear Mixed Model) approach introduced in [72] offered another significant increase in speed. The algorithm is also based on estimating the matrix of pairwise relatedness between individuals, but this matrix is based on a set of SNPs. Also, this matrix is spectrally decomposed in such a way as to transform the SNP data and covariates into independent variables. Due to these adaptations, both the numeric complexity of the algorithm and memory required is linear in the sample size and the number of SNPs. Hence, the algorithm can successfully deal with much larger data sets than the EMMAX algorithm. In addition, the results from the analysis of the WTCCC data set using this algorithm are very similar to the results obtained using the EMMAX algorithm.

The methods described above only consider the application of single marker tests after correcting for the effects of the population structure. A multilocus mixed-model (MLMM) approach is presented in [116]. The authors apply the EMMAX algorithm to model the population structure (although the approach could be integrated with the FaST-LMM algorithm). However, instead of carrying out the appropriate single marker tests based on this correction, they use a stepwise regression procedure. The genetic and error variances are reestimated at each step of the algorithm. Initially, SNPs are introduced into the model (forward regression) until the increase in the proportion of the variance explained is below a (small) threshold or a set number of factors have entered the model. As in the MOSGWA algorithm, a liberal condition is used to build up an overly complex model, which is then pruned using a more stringent condition.

Once the forward regression procedure has been completed, the least significant factor is removed until the selection condition is satisfied. Two criteria are considered:

(i) the mBonf criterion—the procedure ends when all the p-values corresponding to the significance tests for the individual factors are below the Bonferroni threshold.

(ii) the extended BIC criterion—the backward regression procedure ends at the point when removal of a variable would lead to an increase in the value of the BIC_1 function, where $BIC_1 = -2 \ln L_M + k \ln n + 2 \ln[\tau(S_k)]$, L_M is the likelihood of the data under the current model, n is the sample size, k is the number of parameters (SNPs) in the model and $\tau(S_k)$ is the number of possible models with k parameters.

The criterion used here is a particular case of the set of extended BIC criteria, BIC_γ, introduced in [26] where $0 \leq \gamma \leq 1$ and

$$BIC_1 = -2 \ln L_M + k \ln n + 2\gamma \ln[\tau(S_k)].$$

Hence, The BIC_0 criterion is the original BIC criterion, which is suitable for selecting a model when the number of independent variables, m, is much smaller than the sample size n. Under sparsity, it is assumed that the appropriate value of k is much smaller than m. In the context of GWAS, $\tau(S_k) = \binom{m}{k}$. It follows that for $k \ll m$, $\tau(S_k)$ is of order m^k. It can thus be seen that the BIC_1 criterion penalizes large models more than the mBIC2 criterion does and thus seems to be applicable to model selection under extreme sparsity. Indeed, the empirical results given in [116] show that the BIC_1 criterion is more stringent than the mBonf criterion and is thus clearly more adapted to FWER control than to FDR control.

On the other hand, using data from the population in northern Finland, [110], the MLMM procedure detected several more causal SNPs than single marker tests. The results indicate that this multiple-locus approach can take the association between SNPs into account in the testing procedure, unlike single marker tests. At present, there is ongoing research into the application of the mBIC2 criterion. This approach should have greater detection power than the MLMM approach, since the mBIC2 criterion leads to FDR control, rather than FWER control, which is more conservative.

5.5   Admixture Mapping

Using a family-based design, [121] showed that, in addition to information regarding genotypes, information regarding the ancestry of a given portion of the genome can be very useful in association studies based on an admixed population, i.e., one where two previously separated subpopulations become mixed, see Sect. 2.2.4.1. In presently ongoing work, Szulc et al. [120] consider the case when individuals come from such an admixtured population. As noted in [104], when a disease is more prevalent in one subpopulation, then any allele which is more common in that subpopulation is very likely to be inferred to be causal. Denote the possible alleles at a locus by a and A and the ancestral lines as b and B. Let x_{ij} denote the genotype of the ith individual at locus j, where

$$x_{ij} = \begin{cases} -1, & \text{genotype is } a \\ 0, & \text{genotype is } aA. \\ 1, & \text{genotype is } AA \end{cases}$$

Let z_{ij} denote the ancestral origin of locus j for individual j, where

$$z_{ij} = \begin{cases} -1, & \text{origin is } bb \\ 0, & \text{origin is } bB. \\ 1, & \text{origin is } BB \end{cases}$$

where bB means that one allele was inherited from ancestral line b and the other from ancestral line B.

It is assumed that both the genotype and the ancestral origin are known. The inference of genotypes from genome sequencer data was described in Sect. 5.2. Naturally, the ancestral origin of alleles is unknown, but can be inferred to a high level of confidence based on allele frequencies in the ancestral populations and the high level of linkage disequilibrium which characterizes recently admixed populations. The correlation between genotypes decays exponentially in the genetic distance between loci and the time elapsed since the subpopulations came together [24]. For example, [82] notes that the European population shows variation at many sites where only one variant is present in the West African population. Hence, if an individual of mixed race has an allele which is only observed in the European population, then the neighboring positions are almost certainly of European origin. Similarly, Price et al. [99] identify 1,649 markers where one variant can be inferred to be of Latin American origin. Due to such markers being of relatively high density and the high level of linkage disequilibrium, even over long distances, in recently admixed populations we can infer the origin of alleles with near certainty, as long as we have the genetic sequences of individuals from both ancestral subpopulations.

Price et al. [100] describe the HAPMIX algorithm for inferring ancestry, which uses a hidden Markov model to provide a probabilistic estimate of ancestry at each locus. The observations are the genotypes at each locus and the ancestral state is the hidden state to be inferred. Based on the genotype data, HAPMIX estimates the probabilities of the possible phases and calculates the appropriate weighted average of the probabilities that these haplotypes come from a given subpopulation. Using simulations, they conclude that if the subpopulations have only mixed for several generations, then the ancestral origin can be inferred almost with certainty. Even if human subpopulations have mixed for several thousand years, then such inference is reasonably accurate.

Returning to GWAS for admixed populations, let q_i denote the proportion of alleles that individual i has inherited from ancestral line B. Hence,

$$q_i = \frac{1}{2m} \sum_{j=1}^{m} (z_{ij} + 1).$$

Let y_i be the value of a continuous trait for individual i. We can consider the linear regression model

$$y_i = \mu + a_0 q_i + \sum_{j \in G} \beta_j x_{ij} + \sum_{j \in A} \gamma_j z_{ij} + \varepsilon_i,$$

where G is the set of genetic effects, A is the set of ancestral effects and a_0 describes the overall ancestral effect, which prevents spurious effects being inferred when the mean value of the trait is greater in one of the subpopulations than in the other (this may result from environmental factors).

The algorithm for model selection is somewhat similar to the MOSGWA algorithm (see Sect. 5.4.4). First, the individual ancestral effects and the genetic effects are ranked on the basis of a test for the significance of that effect on the basis of a regression model which takes into account only the overall ancestral effect and the effect of interest, i.e.,

$$y_i = \mu + a_0 q_i + \beta_j x_{ij} + \varepsilon_i,$$

in the case of genotype j and

$$y_i = \mu + a_0 q_i + \gamma_j z_{ij} + \varepsilon_i,$$

in the case of ancestral effect j. Using such single marker tests, when correcting the effects of multiple testing, we should take into account the fact that the ancestral state variables corresponding to neighboring sites will be very strongly correlated. Hence, we should use a much more liberal correction factor when assessing the significance of the ancestral state variables. In terms of applying the Bonferroni correction, this means that the effective number of tests corresponding to the ancestral state variables, m^{eff}, is much lower than the actual number of tests, m. This idea is similar to the concepts presented in Sect. 4.1.3 regarding correction for multiple testing in QTL mapping. The details of this calculation are given in [15].

Hence, in order to rank the putative effects, we multiply the p-values associated with the genotype effects by m and the p-values associated with the ancestral effects by m^{eff}. The putative effects are then ranked from the smallest corrected p-value upward. An initial regression model is constructed using a forward step-wise procedure. The variables are considered in the same order as the ranking and a variable is included in the model as long as the value of the standard BIC criterion $BIC = -2 \ln L_M + k \ln n$ decreases, where k is the total number of effects in the model. This forward procedure continues until all the variables with uncorrected p-values smaller than 0.15 have been considered or 80 variables have been included in the model. This forward procedure is based on a liberal criterion, which means that the model obtained will contain too many variables.

The second step involves backward elimination, where the least significant factor is removed from the model as long as the value of the mBIC2 criterion decreases. Hence, a more stringent criterion is used in order to prune the model. Once backward elimination has finished, then variables are added into the model as long as the value of the mBIC2 criterion decreases.

Simulations indicated that this procedure both controlled the FDR and achieved high power in comparison to other methods. The genotypes of 1000 individuals of mixed West African and European descent were simulated based on data from the HapMap Consortium [52] and the hybrid isolation model from [74]. The mean admixing time was taken to be 10 generations and the mean proportion of the genome inherited from West Africa to be 0.7. Almost half a million SNPs were present. Since the ancestral states at loci on different chromosomes are inherited independently of one another, one can obtain the effective number of ancestral sites by adding the effective number of ancestral sites on each chromosome. This gave $m^{eff} = 4722$, which is approximately 1 % of the number of SNPs.

First, the values of a continuous trait were generated from the normal distribution under the assumption that there were no genetic or ancestral effects. The results indicated that the procedure controlled the FWER at below the required level of $\alpha = 0.05$. When just genetic effects were considered in the model, the FWER was 0.01. When both genetic and ancestral effects were considered, the FWER was 0.035.

Second, two models based on 24 causal SNPs were considered. All of these causal SNPs were common (MAF $\geq$ 0.4). Eight of these SNPs were very strongly correlated with some neighboring SNPs (the correlation coefficient between the genotype at the causal SNP and at least one of the 50 nearest neighbors on either side was greater than 0.94) and had the same allelic frequencies in the ancestral populations. Eight of these SNPs were also very strongly correlated with some neighboring SNPs (in the same way), but the allelic frequencies in the ancestral populations differed by at least 0.7. The final group of eight SNPs were weakly correlated with neighbors, but the allelic frequencies in the ancestral populations were again significantly different. According to the first model, there were only genetic effects and these were of equal strength. According to the second model, the effects at 12 loci were genetic (4 loci from each of the three groups of 8) and the effects at the remaining 12 loci were partially genetic and partially ancestral (2/3 genetic, again there were 4 loci from each of the three groups of 8). Overall, the effects of the loci were equal. After generating the trait values based on the appropriate model, the causal SNPs were removed from the data set.

The number of realizations carried out for each model was 250. In each case, the effects were detected in three ways: (i) single marker tests based on the Bonferroni correction aimed at FWER=0.05, (ii) single marker tests based on the Benjamini–Hochberg correction aimed at FDR=0.05, (iii) a multivariate regression procedure using the mBIC2 criterion. Since single marker tests do not take into account the correlation between SNPs, these tests often detected a "clump" of loci surrounding a causal SNP. The effect is almost nonexistent in the case of the multivariate regression procedure. A detection using the mBIC2 criterion was defined to be true when the correlation between the detected genotype and the genotype at the causal SNP was greater than 0.3. Multiple detections of the same causal SNP were counted as just one true positive. All other detections were defined to be false. In the case of the single marker tests, the detection region of a SNP was defined to be the neighborhood of the detected SNP such that the coefficient of correlation with neighboring sites was >0.3. If regions of "clumped" detections intersected, then their detection regions were combined into one group. If such a region covered a causal SNP, then it was defined to be a true discovery. Otherwise, it was defined to be a false positive. The mean numbers of true positives (TP), false positives (FP), and the FDR for these procedures are given in Table 5.7. It should be noted that using the mBIC2 criterion leads to clearly increased power, while controlling the FDR. The Benjamini–Hochberg procedure, on the other hand, does not seem to control the FDR at the required rate.

Table 5.7 Mean number of true positives (out of 24), false positives and the FDR for various procedures for a simulated GWAS based on a admixed population

	Bonferroni	Benjamini–Hochberg	mBIC2
Model 1			
TP	10.71	15.80	21.52
FP	0.12	1.35	0.85
FDR	0.01	0.08	0.04
Model 2			
TP	11.62	15.56	20.38
FP	0.19	1.59	0.90
FDR	0.02	0.09	0.04

It should also be noted that other simulations indicate that simultaneous detection of genetic and ancestral effects resulted in a greater power than combining the results from analyzing the genetic effects and ancestral effects separately. Finally, the simulations were repeated with varying sample sizes from 650 to 1000. The results indicate that for smaller sample sizes ancestral information can be useful in detecting causal SNPs. This is particularly the case when there is a very low level of linkage disequilibrium along a particular section of the chromosome. In fact, in case when the causal SNP is not actually observed and the allele frequencies in the original populations were substantially different, the only reasonably powerful method of detecting that such a causal SNP exists in that region is on the basis of information regarding the ancestry. However, similarly to QTL mapping, the precision in estimating the position of a causal SNP on the basis of ancestral history is very low, due to the very high correlation between the ancestral states at neighboring sites on a chromosome.

The results of these simulations are promising. Since the simulations assumed that the ancestral state was known, it remains to see to what degree inference of the ancestral states lowers the power of these tests. The inference of ancestral states is also clearly a requirement for the analysis of real data.

5.6   Gene–Gene Interaction

5.6.1   Analyzing Gene–Gene Interaction via ANOVA

The classical approach to analyzing gene–gene interaction would be to use analysis of variance (ANOVA). Suppose that Y is a continuous variable and $\alpha_i, \beta_j, i, j = 1, 2, 3$, describe the effects of the three possible genotypes at two different loci. Let 1 denote aa, 2 denote aA and 3 denote AA. Using standard one-way ANOVA, we can model the marginal effect of the first locus on the trait based on the following model:

$$Y = \mu + \alpha_i + \varepsilon,$$

where μ is the mean value of the trait, α_1, α_2 and α_3 describe the effects of the three possible genotypes (measured in relation to the mean value) and ϵ is a random term, which is assumed to have a normal distribution with mean zero and variance independent of the genotype. Since μ is the overall mean, we have $\sum_{i=1}^{3} n_i \alpha_i = 0$, where n_i is the number of individuals with genotype i. It should be noted that the ANOVA procedure is best adapted to experimental settings in which the group sizes are equal, i.e., $n_1 = n_2 = n_3$. In this case, we have $\alpha_1 + \alpha_2 + \alpha_3 = 0$. In order to test whether this locus is associated with the trait of interest, we test the null hypothesis of no association, $H_0 : \alpha_1 = \alpha_2 = \alpha_3 = 0$, against the alternative that not all of these parameters are equal to zero. The details of the testing procedure are given in Sect. 8.3.1.

In order to model how the trait depends on both genotypes, we can model the effect of two loci based on the following model:

$$Y = \mu + \alpha_i + \beta_j + \delta_{ij} + \varepsilon,$$

where α_1, α_2 and α_3 give the individual effects of the three possible genotypes at the first locus, β_1, β_2 and β_3 the individual effects of the three possible genotypes at the second locus and the set of $\delta_{ij}, 1 \le i, j \le 3$ describe the interactions between the nine possible pairs of genotypes. For example, when $\delta_{11} = 2$, then the mean value of the trait for individuals with genotype aa is increased by 2 in comparison to the mean value when only the individual effects α_1 and β_1 are taken into account. In this case, the overall mean of the trait is μ, the mean for those with genotype i at locus 1 is $\mu + \alpha_i$ and the mean for those with genotype j at locus 2 is $\mu + \beta_j$. Hence,

$$\sum_{i=1}^{3} n_{i\bullet}\alpha_i = \sum_{j=1}^{3} n_{\bullet j}\beta_j = 0,$$

$$\sum_{i=1}^{3} n_{ij}\delta_{ij} = 0, \quad j = 1, 2, 3$$

$$\sum_{j=1}^{3} n_{ij}\delta_{ij} = 0, \quad i = 1, 2, 3,$$

where $n_{i\bullet}$ is the number of individuals with genotype i at locus 1, $n_{\bullet j}$ is the number of individuals with genotype j at locus 2 and n_{ij} is the number of individuals with the pair of genotypes (i, j). In this case, we test the null hypothesis that these loci are not associated with the trait, $H_0 : \alpha_i = 0, \beta_j = 0, \delta_{ij} = 0$, for all $1 \le i, j \le 3$ against the alternative that at least one of the parameters is not equal to zero.

In addition, we may consider the following model in which there are just individual effects, i.e., without any interaction

$$Y = \mu + \alpha_i + \beta_j + \varepsilon.$$

In this case, we test the null hypothesis that these loci are not associated with the trait, $H_0 : \alpha_i = 0, \beta_j = 0$, for all $1 \leq i, j \leq 3$ against the alternative that at least one of the parameters is not equal to zero. The details of the testing procedures for these two tests are described in Sect. 8.3.2.

In the case where such a continuous trait does not have a normal distribution, then the appropriate nonparametric test, the Kruskal–Wallis test, can be applied, see Sect. 8.7. When the trait of interest is dichotomous, we can code it using a 0–1 variable and use the logistic transformation $Y = \ln[p/(1 - p)]$, where p is the proportion of individuals in a group with trait value 1.

One can also extend these models to consider a larger number of loci. A major problem of such an approach lies in the number of parameters which need to be estimated. One-way ANOVA based on one locus requires the estimation of two parameters, α_1 and α_2 (since $\sum_{i=1}^{3} n_i \alpha_i = 0$). Two-way ANOVA with interaction requires the estimation of eight parameters α_i, β_j and δ_{ij} for $i, j \in \{1, 2\}$, (since the remaining parameters can be calculated from the appropriate conditions). In general, in order to model all the possible individual effects and interactions of k loci, one would have to consider $3^k - 1$ parameters. One could use appropriate methods of model selection to decide which of these parameters are significantly different from zero. However, given that there are m SNPs, where m is very large, there are $\binom{m}{k}$ ways of choosing k loci, this would involve a sequential approach to model selection.

More importantly, the number of sets of k loci is of order m^k and so any systematic investigation of all the possible interactions would involve excessive computation time. For example, Marchini et al. [77] state that it took 33 h on a 10-node cluster to carry out an analysis of the all the pairs from a set of 300,000 loci for 1,000 cases and 1,000 controls. Since there are approximately 100,000 times as many triplets as pairs in this case, to carry out a similar analysis of all triplets would take more than 100,000 times as long, given that the analysis of a triplet is more complex than the analysis of a pair. In his review of research on GWAS, Cordell [33] notes the exponential increase in time required to systematically analyze interactions between k loci. He also states that multiple testing procedures would have to be adapted to take linkage disequilibrium into account. Otherwise, such procedures would naturally be far too conservative. De et al. [35] use initial filters based on single marker tests and knowledge regarding genes that are likely to interact together to reduce the search space. Emily et al. [42] also use an approach based on biological networks to reduce the search space when looking for interactions. However, Moore [86] states that the use of single marker tests to reduce the search space is problematic due to fact that many complex traits are subject to epistatic effects (i.e., interaction) rather than the individual effects of genes, thus single marker tests will very often fail to detect genes that are involved in interactions.

Table 5.8 Analysis of the effect of a pair of loci using multifactor dimensionality reduction

	aa	aA	AA
aa	**(40, 20)**	(40, 80)	**(40, 20)**
aA	(40, 80)	**(160, 80)**	(40, 80)
AA	**(40,20)**	(40, 80)	**(40,20)**

The data before dimensionality reduction. In each cell, the first number gives the frequency of cases, the second gives the frequency of controls. High risk genotypes are highlighted in bold. The row corresponds to locus 1, the column to locus 2

Table 5.9 Analysis of the effect of a pair of loci using multifactor dimensionality reduction

	High risk	Low risk
Frequencies	(320, 160)	(160, 320)

The data after dimensionality reduction. In both cells, the first number gives the frequency of cases, the second gives the frequency of controls

5.6.2   Multifactor Dimensionality Reduction

The approach of multifactor dimensionality reduction (MDR) [107] deals with the problem of parameter estimation, but requires adaptation to deal with the problem of having to search a huge space to detect complex interactions and to deal with continuous traits. Suppose that the trait of interest is dichotomic. Using the initial formulation, a set of k genotypes is reduced to a 0–1 variable, where 0 denotes low-risk groups and 1 denotes high risk groups. In its simplest form, the number of cases is equal to the number of controls. Consider the genotypes at two loci and suppose that the number of cases and controls according to the pair of genotypes observed is given in Table 5.8. A pair of genotypes is defined to be high risk when the number of cases with that genotype is greater than the number of controls with that genotype.

This approach can be adapted to studies in which the number of controls differs from the number of cases. In this case, a pair of genotypes is defined to be high risk when the proportion of cases among the individuals with that pair of genotypes is higher than the proportion of cases among all individuals. However, the method is most effective when the frequency of cases and controls are very similar. Application of MDR reduces this table to Table 5.9. This is a two by two contingency table, but standard tests of association should not be used to test the hypothesis of no association, as such tests naturally underestimate the appropriate p-value for the test, due to the procedure maximizing the difference between the high-risk and low-risk groups.

One can test the hypothesis of no association based on the consistency of the models based upon 90 % of the data, i.e., the proportion of times this procedure gives us the same model. If there is a significant association between the pair of genotypes and the dichotomous trait, then the model based on a proportion of the data should give the same split into high-risk and low-risk groups a very large proportion of times.

One can obtain an estimate of the distribution of this statistic under the assumption of no association by e.g.

1. Using the permutation approach, simulate 1,000 different samples of the same size as the study sample. In each case split the data into high-risk and low-risk groups.
2. For each sample, form 1,000 subsamples in which 10 % of the original data are removed at random. Split each reduced data set into high-risk and low-risk groups. Calculate the proportion of the subsamples for which the split into high-risk and low-risk groups is the same as in the original sample.
3. Based on the first two steps, we have 1,000 proportions, which can be used to estimate the distribution of the test statistic.

It is very easy to generalize this approach to studying the effect of k rather than 2 genotypes. In GWAS, the increase in the numerical complexity of such analysis is outweighed by the number of sets of k loci that exist, which thus remains a major limiting factor in analyzing gene–gene interaction. MDR has been very successful in the analysis of factors in sporadic cancers, see e.g. [4]. However, such studies use a small number of genotypes compared to GWAS, since they are based on knowledge of genes that are e.g. responsible for DNA replication and DNA repair. Thus the search space is already reduced greatly. On one hand, the basic MDR algorithm is highly flexible, since there are no assumptions regarding the underlying model. On the other hand, this "agnosticism" is problematic, since without further information it is impossible to define an effective search algorithm for large-scale problems. This is particularly true if epistatic effects are the most important. In this case, filtering out nonsignificant single markers will be an ineffective approach to model selection. Another problem lies in the fact that it might well be difficult to interpret the split of genotypes into high risk and low risk.

Another problem associated with investigating high-order interactions is the number of possible combinations of genotypes when k loci are considered. The number of such combinations is 3^k. In GWAS studies, there are often around 1,000 cases and controls. Hence, for $k = 4$, on average there will be around 24 individuals in total with each combination of genotypes. For $k = 5$, this number will be approximately 8, a rather small number for investigating what might be a fairly simple genetic interaction in real terms. On the other hand, Riveros et al. [108] note that graphs based on discovered pairwise interactions can also give information on interactions between a larger number of loci. In simplistic terms, if there exist interactions between the following three pairs: (A, B), (A, C) and (B, C), then it seems reasonable that the three loci A, B and C interact.

A number of important adaptations have been made to MDR. Pattin et al. [93] use the fact that MDR maximizes the difference between the high-risk and low-risk groups to derive a test whose statistic has an extreme value distribution. Lou et al. [75] adapted MDR to problems where the dependent variable is continuous rather than dichotomous. Gui et al. [51] describe a method of implementing this approach when the search space is large. A number of ways for making search more effective in practice have been considered, see [34, 35, 42]. These methods are based on available knowledge and/or first using single marker tests (which, as described above, has its limitations).

5.6.3 Logic Regression in GWAS

The general concepts of logic regression were presented in Sect. 4.4 and so here we just present the application of logic regression to detecting gene–gene interactions in GWAS. Although the problem of model selection in QTL mapping can be already fairly computationally difficult, the number of SNPs considered in GWAS is generally very much larger and the number of possible interactions increases exponentially. Hence, the main differences in the application of logic regression to GWAS result from the huge dimension of the search space.

Kooperberg and Ruczinski [65] see logic regression as a very useful tool in discovering possible interaction effects, which should then be investigated further. Each gene can have a dominance effect and/or a recessive effect. The dominance effect of allele a at the ith SNP is associated with the covariate $X_{i,1}$, where $X_{i,1} = 1$ if the genotype at this locus is aa or aA and $X_{i,1} = 0$ if the genotype at this locus is AA. The recessive effect of allele a at the ith SNP is associated with the variable $X_{i,2}$, where $X_{i,2} = 1$ if the genotype at this locus is aa and $X_{i,1} = 0$ if the genotype at this locus is aA or AA.

They argue that by just developing one regression model in a GWAS study, one ignores the fact that the variables in the model may well not be causal, but simply associated with causal variables. Also, the power of detecting interactions is likely to be very small. Hence, they use an approach based on simulated annealing in order to generate a large number of models. The SNPs that appear most commonly in these models can then be treated as targets for further study. Simulated annealing creates a chain of models, where each model is a modification of the previous model. Suppose at a given stage the model for a continuous trait Y is

$$Y = \beta_0 + \sum_{i=1}^{k} \beta_i L_i,$$

where L_i is a logical expression. A proposition for the next model of the chain is obtained by modifying one of the logical expressions in L_i. Suppose the logical expression L_i is $X_{j,1} \wedge (X_{k,2} \vee X_{l,1})$, which is equal to one if the genotype at the jth locus is aa or aA and either the genotype at the kth locus is aa or the genotype at the kth locus is aa or aA, otherwise it is equal to zero. Such an expression implies an interaction between the jth SNP and the other two. There are various ways in which this expression can be modified to obtain the proposition for a new model. For example, one of the decision variables could be altered. Hence, changing the first variable, we obtain $X_{j',1} \wedge (X_{k,2} \vee X_{l,1})$. In addition, a logical operation might be added, removed, or changed. The regression model obtained in this way can be given

a score based on Akaike's Information Criterion (AIC). The number of variables
in this model is defined to be the total number of genetic variables contained in
the logical expressions $L_1, \ldots, L_k$. It is noted that changing the coefficient of the
model size in AIC corresponds to changing the expected size of a model when the
prior distribution of model size is geometric [54]. The proposed model is always
accepted when its score according to AIC is better (lower) than the score of the
present model. When the proposed model has a worse score than the present model,
then the proposed model is accepted with some probability which is decreasing in
the difference between the scores (i.e., the worse the proposal, the less likely it is to
be accepted). Also, the probability of accepting a worse model falls as time goes on.
If a proposition is rejected, then the present model is accepted as the new model and
another modification will be attempted. Hence, using simulating annealing, initially
the scores of the visited models will only have a slight tendency to improve on
average and a wide range of models will be generated. However, as time goes on,
only improved models are likely to be selected and eventually the algorithm reaches a
model which has a better score than all the models in the surrounding neighborhood.

Various statistics regarding the models generated in this chain are used to interpret
the results from this chain:

1. The distribution of the size of these models. It should be noted that these models
   will on average tend to be overcomplicated. One should compare the distribution
   of the size of the model with the distribution of the size of the model based on
   random permutations of the trait values (which leads to the trait being independent
   of the genetic code). If these distributions are significantly different, then we have
   evidence of genetic effects.
2. The fraction of models that contain a particular covariate. Covariates which appear
   in many models are likely to correspond to causal SNPs or to loci which are very
   strongly linked to causal SNPs.
3. The fraction of models that include pairs, triplets and quartets of SNPs in the
   same logical expression. If the frequency of the appearance of a pair of SNPs in
   the same logical expression is greater than the product of the frequencies with
   which the individual SNPs appear, this is an indication of a possible interaction
   between the two corresponding loci.

Using different coefficients in the AIC (i.e., different priors on the size of the
model) does lead to significant differences in the sizes of the models obtained, but
the effect on the rankings of the frequencies with which covariates appear in the
model is almost unchanged. Results obtained from the analysis of real data indicate
that this approach is very useful.

Chen et al. [25] give a review of the application of logic programming to GWAS.
One interesting development of the method proposed by Kooperberg and Ruczinski
[65] is the Logic Feature Selection (LogicFS) approach [114]. This involves the
resampling of data and subsequent fitting of models using logic regression. This
approach also puts logical expressions into a standard form (disjunctive normal form),
which makes it easier to interpret and compare models. Schwender et al. [115] further

adapt this model by grouping neighboring sites to counteract the lack of power of such tests resulting from strong linkage disequilibrium and the small effect of individual sites. This approach is similar to that described in Sect. 5.3.5. Wolf et al. [125] apply a resampling procedure to generate a set of logic regression models, which they call a logic forest. Simulations indicate that the results obtained using this approach are comparable to those obtained using Logic Regression and more robust than those obtained using the original approach based on simulated annealing.

5.7 Other Recent Advances and the Outlook for GWAS

In this section, we will look at the problem of determining genetic factors for a common disease, i.e., the trait of interest will be dichotomous. Due to the huge amount of data that are already available from genome sequencers and the fact that the genetic effects underlying common diseases are complex, efficient methods are needed to analyze GWAS. Around 100 million SNPs have already been found and validated in the human genome [89]. Even with the huge increase in computing speeds and memory capacity, it is necessary to select SNPs for analysis. First, we can eliminate SNPs where the minor allele frequency is low in both the case group and study group. Second, we can use biological information to further reduce the set of SNPs. One such way is to only consider SNPs where the two alleles result in the production of different proteins (i.e., missence mutations [57]). Additionally, one can use tag SNPs which are in strong linkage disequilibrium with other loci which are not included in the analysis. For approaches to choosing a set of tag SNPs, see [21, 46]. de Bakker et al. [36] find that using tag SNPs which are strongly correlated with all the other SNPs in the group they represent ($r^2 > 0.8$) has very little effect on the power of detecting main effects. Using such screening, the number of SNPs included in the analysis can be reduced to the order of 10,000. Data sets of this size enable the search for interactions between SNPs.

Chen et al. [29] carry out a comparison of eight methods using logistic regression as the baseline for comparison.

1. Logistic regression with main effects only (LR).
2. Logistic regression with only interaction terms, which are represented by the products of variables corresponding to the interacting SNPs.
3. Full interaction model (FIM—[77]). This is based on logistic regression, but when analyzing possible interactions allows each of the nine possible genotype pairs to be associated with different risk levels.
4. SNP Harvester (SH—[127]). This is based on logistic regression. A model is built based on a random subset of SNPs and this model is then improved using a process similar to simulated annealing. This process is repeated. On the basis of these procedures, a ranking of kth order interactions can be made, for $1 \le k \le \log_3 N_d - 1$, where N_d is the size of the case group, typically $1 \le k \le 5$. Chi-square tests are then used to test the significance of an effect/interaction.

5. Multifactor dimensionality reduction (MDR—[107]).
6. Information gain (IG—see [87]). This is based on a ranking of pairs of SNPs according to the gain in information gained by considering two SNPs together rather than separately. The information measure is based on Shannon information (see [94]).
7. Bayesian Epistasis Association Mapping (BEAM—see [132]). This is based on a Bayesian approach where SNPs have prior probabilities of being in one of three classes: (a) noncausal, (b) marginal effect, (c) interactive effect. Markov Chain Monte Carlo methods are applied to obtain the posterior probabilities of being in each group. Likelihood ratios are used to estimate p-values to test for the appropriate effects.
8. Maximum Entropy Conditional Probability Modeling (MECPM—[85]). This approach builds up a model of the conditional distribution of the dichotomous trait of interest step by step by maximizing Shannon's information measure. Up to five-way interactions can be modeled. The Bayesian Information Criterion (BIC) is used to decide when to stop building the model. This approach is similar in many ways to the concept of Bayesian networks, which are used to describe the interrelationships between discrete variables [45].

In many ways, it is difficult to compare these methods. Some of them are based on testing procedures which state whether given interactions are significant based on p-values and a multiple testing procedure, e.g., FIM, BEAM, and SH. MECPM does not give p-values for individual effects, but uses the BIC criterion to decide which effects are significant. Other procedures simply produce a ranking of effects. However, ordering p-values gives us a ranking and the MECPM procedure also introduces effects in order of "their significance." Hence, Chen et al. [29] compare the results obtained on the basis of the top ranked effects of each order. They use data sets which contain between 1,000 and 10,000 SNPs and between 500 and 4,000 individuals in both the case group and the control group. The programs, all written by the original authors of the procedures, were run using a Windows Operating System with 3G of CPU and 2G of RAM. The default settings were used for each of the programs, except in the case of MDR. This was due to the lack of memory space available. The default setting for MDR is to carry out a full search of the set of possible interactions (here assumed to be up to fifth order). However, when there were 1,000 SNPs only the simpler heuristic search procedure could be run given the memory capacity and neither procedure would run when there were 10,000 SNPs.

Three test scenarios were used to compare these procedures:

1. The null model. The condition in question is not associated with any SNP.
2. A model based on interactions. In this case there were five interactions. Interactions 1 and 2 involved two loci, interactions 3 and 4 involved three loci and interaction 5 involved five loci. Interaction 1 was based on a slightly increased risk when the major allele was present at both sites. The frequency of this allele was 0.75 at both sites. In this case, the odds ratio for the main effects for both loci was 1.15. The odds ratio for the interaction effect was 1.16. Hence, a large

proportion of individuals have a slightly increased risk based on these two loci. Interaction 2 was based on a large increase in risk when the minor allele was present at both sites, especially if at least one of the sites was homozygous. The MAFs at these sites were 0.2 and 0.3, respectively. In this case, the odds ratios for the main effects at these loci were 1.89 and 1.56, respectively (moderately large). The odds ratio for the interaction effect was 3.79 (large). Interaction 3 was based on a large increase in risk when the minor allele was present at all sites, especially if at least two of the sites were homozygous. The MAFs at these sites were 0.4, 0.25 and 0.25, respectively. In this case, the odds ratio for the main effects are 1.16, 1.25 and 1.25 (relatively small effects), respectively and the odds ratio for the interaction was 2.28 (moderately large). Interaction 4 was based on a large increase in risk when an individual has two copies of the minor allele at the first site and at least one minor allele at either of the other sites. The MAFs are 0.25, 0.2 and 0. 2, respectively. In this case, the main effect of A is quite large (odds ratio 2.45), but the main effects of the other two sites are very small (odds ratio 1.06). The odds ratio for the interaction between the three sites is large, 5.79. Interaction 5 involves a large increase in risk when the minor allele is present. The MAF at each site is 0.3. In this case, each of the main effects are small (odds ratio 1.09), while the interaction effect is large (odds ratio, 4.48).
3. A model based on purely marginal effects with 5 causal SNPs. At the first SNP, risk is slightly increased when there is at least one major allele, which has frequency 0.75. At the second SNP, risk is increasing in the number of minor alleles and the MAF is 0.3. The effect is similar, but the risk lower, at the third SNP, where the MAF is 0.4. At the fourth SNP, there is a very high risk when an individual has two copies of the minor allele and the MAF is 0.25. At the fifth SNP, there is a slightly increased risk when an individual has at least one copy of the minor allele and the MAF is 0.3.

The null model was used to test the accuracy of the p-values given by the FIM, BEAM, and SH procedures and the appropriateness of the multiple testing procedures used. It should be noted that the p-values given by the MDR procedure are accurate, since they are based on permutation tests and not on theoretical approximations. However, only the top rank interactions are assigned a p-value, thus the procedure does not automatically define whether a result is significant or not on the basis of a multiple testing procedure. The FDR was controlled at a nominal rate of 10 %. Under the null hypothesis, each of FIM, BEAM, and SH control the FDR, as hence FWER, at the appropriate level. However, it is noticeable that false detections of interactions are very rare. In particular, BEAM does not make any false detections of two- or three-way interactions in 1,000 simulations. This indicates that these procedures might lack power to detect interactions. Although one can estimate the FWER of the MECPM procedure, since it uses a model selection procedure, this was not carried out in [29]. However, the results from the other simulations indicate that the BIC procedure chooses a model of an appropriate size.

Since some of the procedures only rank the SNPs, power is described as a function of K, where K is the number of top ranked sites (or interactions, as appropriate) assumed to be significant, where $K \leq 100$. Obviously, power is a nondecreasing function of K. In the case of MECPM, the procedure only ranked the sites which entered the model and at most 20 sites were included in the model. Hence, for this procedure power curves were given for $K \leq 20$.

The results for the second model were assessed from two points of view on the basis of 100 simulations. First, the power to progressively detect interactions. Overall power was defined to be the proportion of the fifteen causal SNPs that appeared in the first K positions of the ranking. One can define the power to progressively detect a given interaction in a similar way. The power to detect interaction i was the proportion of sites involved in that interaction which were in the first K positions of the ranking. For $K \leq 20$, the MECPM procedure clearly has the greatest overall power and only the SH procedure has a greater power based on the 100 highest ranked sites (obviously at the cost of a large FDR). Based on a case group and control group of size 1,000, regardless of the number of SNPs in the data set, MECPM progressively detects interaction 2 and 3 with power 1, interaction 4 is detected with power above 0.8. Regardless of the method used, interaction 5 is detected very rarely, especially when the number of SNPs considered is increased. Interaction 1 is the most difficult to detect. The SH procedure is ranked second on the basis of this power measure. It discovers each of the interactions with relatively high power, except interaction 4.

The second measure of power was the power to precisely detect interactions. This was based on the K most highly ranked interactions of the appropriate order, apart from the case of MECPM where it was based on the interactions in the final model. MECPM correctly determined interactions 2, 3 and 5 in 96, 94 and 46 % of the simulations. The other two interactions were never precisely detected. As expected from the simulations based on the null model, BEAM is not at all successful in precisely detecting interactions. The power of MDR to precisely detect any of the interactions was very low and the other methods only precisely detected interaction 2 with large power. It should be noted that the IG, LRIT and FIM procedures only detect interactions of order two and so by definition cannot precisely detect interactions 3–5.

The third model was used to assess the power of the methods to detect marginal effects. The IG and LRIT procedures were not used, since they only detect interactions. Apart from MDR, which had clearly the lowest power, the remaining procedures are slight improvements on a standard logistic regression procedure. The SH procedure seems to be marginally better at detecting marginal effects than the other procedures.

Based on this comparison, Maximum Entropy Conditional Probability Modeling (MECPM) seems to be a very natural approach to analyzing GWAS when the trait of interest is dichotomous. One reason for this is that this approach is specifically designed to analyze the joint distribution of a set of discrete variables, where each variable can only take a small number of different values (here the dependent variable is dichotomous and the genetic variables each take one of three possible values). MECPM makes no assumptions regarding the relationship between variables, which seems to cause a small loss in power when an additive model is appropriate in comparison to the SH procedure. However, its overall flexibility makes it very competitive when interactions are taken into account. Another advantage of this approach is that the BIC criterion is used to choose an appropriate model and this seems to give models of an appropriate size. If required, in exploratory studies one could lower the penalty on model size to obtain a larger set of loci for further investigation.

One problem of using a multiple testing approach to detecting interaction when the trait of interest is dichotomous lies in the fact that the data (and hence the distributions of the test statistics) are discrete. This becomes of great importance when the number of tests is very large, since it may become virtually impossible to reject the null hypothesis of no interaction, particularly when we realize that correlation between the test statistics due to linkage disequilibrium will lead to conservative results. The simulations described above indicate that such procedures are indeed conservative. The implementation of MECPM goes a long way toward solving this problem. On the other hand, when the trait of interest is quantitative methods based on regression models such as MOSGWA seem much more natural.

Another problem lies in the fact that any presently used method finds interactions difficult to detect when the marginal effects are small. For example, even though interaction 4 involves a large interaction effect between three sites, it was very difficult to detect its nature. This is because the computing power necessary to carry out exhaustive search of interactions up to a certain level is simply too high (the MDR procedure crashed when exhaustive search was attempted). For this reason, MECPM uses a form of greedy search in which individual SNPs are added sequentially to the model, either as marginal effects or interacting with SNPs which are already in the model. For example, when a strong three-way interaction is present where the marginal effects of at least two of the alleles are weak, the MECPM procedure almost certainly will not be able to accurately describe this interaction. This is due to the fact that it uses a sort of one-step ahead search procedure and in order to obtain a better model it is necessary to add two SNPs at once. Exhaustive search for such interactions would require a huge increase in computational power and even heuristics involving a two-step-ahead search procedure without information on what interactions are likely would involve a large increase in computational power. In such a scenario, it seems that the way forward involves expert information which leads to improved search heuristics.

The model underlying MECPM is clearly different from the models previously described in this book. However, the criterion used for model selection is strongly embedded in the ideas running throughout this book. This illustrates that cross fertilization of ideas leads to advances in science. One thread that has run through this book is that very different statistical philosophies and approaches lead to results which are remarkably similar. This often leads to a deeper mathematical understanding of both the world of concepts and the world of nature.

Finally, one might think that increasing computational power might eventually lead to the possibility of exhaustive search procedures. However, it is likely that this increase in computational power will be accompanied by an increased amount of data. This will both lead to increased understanding of genetic effects, but also a continued need for efficient algorithms.

References

1. Affymetrix, Inc.: BRLMM: an Improved Genotype Calling Method for the GeneChip Human Mapping 500K Array Set. http://www.affymetrix.com/support/technical/whitepapers/brlmm_whitepaper.pdf (2006)
2. Alexander, D.H., Lange, K.: Enhancements to the ADMIXTURE algorithm for individual ancestry estimation. BMC Bioinform. **12**, 246 (2011)
3. Alexander, D., Novembre, J., Lange, K.: Fast model-based estimation of ancestry in unrelated individuals. Genome Res. **19**, 1655–1664 (2009)
4. Andrew, A.S., Nelson, H.H., Kelsey, K.T., et al.: Concordance of multiple analytical approaches demonstrates a complex relationship between DNA repair gene SNPs, smoking and bladder cancer susceptibility. Carcinogenesis **27**(5), 1030–1037 (2006)
5. Asimit, J., Zeggini, E.: Rare variant association analysis methods for complex traits. Annu. Rev. Genet. **44**, 293–308 (2010)
6. Armitage, P.: Tests for linear trends in proportions and frequencies. Biometrics **11**(3), 375–386 (1955)
7. Balding, D.J.: A tutorial on statistical methods for population association studies. Nat. Rev. Gen. **7**, 781–791 (2006)
8. de Bakker, P.I., Yelensky, R., Pe'er, I., Gabriel, S.B., Daly, M.J., Altshuler, D.: Efficiency and power in genetic association studies. Nat. Genet. **37**, 1217–1223 (2005)
9. Bansal, V., Libiger, O., Torkamani, A., Schork, N.J.: Statistical analysis strategies for association studies involving rare variants. Nat. Rev. Genet. **11**(11), 773–785 (2010)
10. Barlow, R.E., Bartholomew, D.J., Bremner, J.M., Brunk, H.D.: Statistical Inference under Order Restrictions; the Theory and Application of Isotonic Regression. Wiley, New York (1972)
11. Barrett, J.C., Fry, B., Maller, J., Daly, M.J.: Haploview: analysis and visualization of LD and haplotype maps. Bioinformatics **21**, 263–265 (2005)
12. Bazaraa, M., Shetty, C.: Nonlinear Programming: Theory and Algorithms. Wiley, New York (1979)
13. Beben, B., Visscher, P.M., McRae, A.F.: Family-based genome-wide association studies. Pharmacogenomics **20**(2), 181–190 (2009)
14. Benjamini, Y., Hochberg, Y.: Controlling the false discovery rate: a practical and powerful approach to multiple testing. J. Roy. Statist. Soc. Ser. B **57**, 289–300 (1995)
15. Bogdan, M., Frommlet, F., Biecek, P., Cheng, R., Ghosh, J.K., Doerge, R.W.: Extending the modified Bayesian Information Criterion (mBIC) to dense markers and multiple interval mapping. Biometrics **64**, 1162–1169 (2008)
16. Bogdan, M., Żak-Szatkowska, M., Ghosh, J.K.: Selecting explanatory variables with the modified version of Bayesian Information Criterion. Qual. Reliab. Eng. Int. **24**, 627–641 (2008)
17. Browning, S.R.: Missing data imputation and haplotype phase inference for genome-wide association studies. Hum. Genet. **124**, 439–450 (2008)
18. Browning, B.L., Yu, Z.: Simultaneous genotype calling and haplotype phase inference improves genotype accuracy and reduces false positive associations for genome-wide association studies. Am. J. Hum. Genet. **85**, 847–861 (2009)
19. Browning, B.L., Browning, S.R.: A unified approach to genotype imputation and haplotype phase inference for large data sets of trios and unrelated individuals. Am. J. Hum. Genet. **84**, 210–223 (2009)
20. Cantor, R.M., Lange, K., Sinsheimer, J.S.: Prioritizing GWAS results: A review of statistical methods and recommendations for their application. Am. J. Hum. Genet. **86**(1), 6–22 (2010)
21. Carlson, C.S., Eberle, M.A., Rieder, M.J., Yi, Q., Kruglyak, L., Nickerson, D.A.: Selecting a maximally informative set of single-nucleotide polymorphisms for association analyses using linkage disequilibrium. Am. J. Hum. Genet. **74**(1), 106–120 (2004)
22. Carvalho, B., Bengtsson, H., Speed, T.P., Irizarry, R.A.: Exploration, normalization, and genotype calls of high-density oligonucleotide SNP array data. Biostatistics **8**, 485–499 (2007)

23. Carvalho, B.S., Irizarry, R.A.: A framework for oligonucleotide microarray preprocessing. Bioinformatics **26**, 2363–2367 (2010)
24. Chakraborty, R., Weiss, K.M.: Admixture as a tool for finding linked genes and detecting that difference from allelic association between loci. Proc. Nat. Acad. Sci. **85**(23), 9119–9123 (1988)
25. Chen, C.C.M., Schwender, H., Keith, J., Nunkesser, R., Mengersen, K., Macrossan, P.: Methods for identifying SNP interactions: a review on variations of logic regression, random forest and Bayesian logistic regression. IEEE/ACM Trans. Comput. Biol. Bioinf. **8**(6), 1580–1591 (2011)
26. Chen, J., Chen, Z.: Extended Bayesian Information criteria for model selection with large model spaces. Biometrika **95**(3), 759–771 (2008)
27. Chen, J., Chen, Z.: Extended BIC for small n-large-P sparse GLM. www.stat.nus.edu.sg/~stachenz/ChenChen.pdf (2010)
28. Chen, J., Chen, Z.: Tournament screening cum EBIC for feature selection with high-dimensional feature spaces. Sci. China A: Math. **52**(6), 1327–1341 (2009)
29. Chen, L., Yu, G., Langefeld, C.D., et al.: Comparative analysis of methods for detecting interacting loci. BMC Genomics **12**(1), 344 (2011)
30. Chipman, H., George, E.I., McCulloch, R.E.: The practical implementation of Bayesian model selection (with discussion). In: Lahiri, P. (ed.) Model Selection, pp. 66–134. IMS, Beachwood, OH (2001)
31. Colditz, G.A., Hankinson, S.E.: The nurses' health study: lifestyle and health among women. Nat. Rev. Cancer **5**, 388–396 (2005)
32. Consortium WTCCC: Genome-wide association study of 14,000 cases of seven common diseases and 3,000 shared controls. Nature **447**, 661–678 (2007)
33. Cordell, H.J.: Detecting gene-gene interactions that underlie human diseases. Nat. Rev. Genet. **10**(6), 392–404 (2009)
34. Dai, H., Bhandary, M., Becker, M., Leeder, J.S., Gaedigk, R., Motsinger-Reif, A.A.: Global tests of p-values for multifactor dimensionality reduction models in selection of optimal number of target genes biodata mining **5**(1), 1–17 (2012)
35. De, R., Verma, S.S., Holmes, M.V. et al.: Dissecting the obesity disease landscape: identifying gene-gene interactions that are highly associated with body mass index. In: 2014 8th International Conference on Systems Biology (ISB), 124–131. IEEE (2014)
36. de Bakker, P.I., Yelensky, R., Pe'er, I., Gabriel, S.B., Daly, M.J., Altshuler, D.: Efficiency and power in genetic association studies. Nat. Genet. **37**(11), 1217–1223 (2005)
37. Devlin, B., Roeder, K.: Genomic control for association studies. Biometrics **55**, 997–1004 (1999)
38. Di, X., Matsuzaki, H., Webster, T.A., Hubbell, E., Liu, G., Dong, S., Bartell, D., Huang, J., Chiles, R., Yang, G., Shen, M., Kulp, D., Kennedy, G.C., Mei, R., Jones, K.W., Cawley, S.: Dynamic model based algorithms for screening and genotyping over 100K SNPs on oligonucleotide microarrays. Bioinformatics **21**, 1958–1963 (2005)
39. Dolejsi, E., Bodenstorfer, B., Frommlet, F.: Analyzing genome-wide association studies with an FDR controlling modification of the Bayesian Information Criterion. PLoS One e103322 (2014)
40. Dudbridge, F., Gusnanto, A.: Estimation of significance thresholds for genomewide association scans. Genet. Epid. **32**, 227–234 (2008)
41. Eichler, E.E., et al.: Missing heritability and strategies for finding the underlying causes of complex disease. Nat. Rev. Genet. **11**, 446–450 (2010)
42. Emily, M., Mailund, T., Hein, J., Schauser, L., Schierup, M.H.: Using biological networks to search for interacting loci in genome-wide association studies. Eur. J. Hum. Genet. **17**(10), 1231–1240 (2009)
43. Fan, J., Lv, J.: Sure independence screening for ultrahigh dimensional feature space. J. R. Statist. Soc. B **70**, 849–911 (2008)
44. Freidlin, B., Zheng, G., Li, Z., Gastwirth, J.L.: Trend tests for case-control studies of genetic markers: power, sample size and robustness. Hum. Hered. **53**, 146–152 (2002)

45. Friedman, N., Linial, M., Nachman, I., Pe'er, D.: Using Bayesian networks to analyze expression data. J. Comput. Biol. **7**(3–4), 601–620 (2000)
46. Frommlet, F.: Tag SNP selection based on clustering according to dominant sets found using replicator dynamics. Adv. Data Anal. Classif. **4**, 65–83 (2010)
47. Frommlet, F., Chakrabarti, A., Murawska, M., Bogdan, M.: Asymptotic Bayes optimality under sparsity of selection rules for general priors. arXiv:1005.4753 (2010)
48. Frommlet, F., Ruhaltinger, F., Twarog, P., Bogdan, M.: Modified versions of Bayesian information criterion for genome-wide association studies. CSDA **56**, 1038–1051 (2012)
49. George, E.I., Foster, D.P.: Calibration and empirical Bayes variable selection. Biometrika **87**, 731–747 (2000)
50. Griffin, J.E., Brown, P.J.: Bayesian adaptive lasso with non-convex penalization. Technical Report, University of Kent (2007)
51. Gui, J., Moore, J.H., Williams, S.M., Andrews, P., Hillege, H.L., van der Harst, P., Navis, G., Van Gilst, W.H., Asselbergs, F.W., Gilbert-Diamond, D.: A simple and computationally efficient approach to multifactor dimensionality reduction analysis of gene-gene interactions for quantitative traits. PLoS One **8**(6), e66545 (2013)
52. Nature Consortium.: A second generation human haplotype map of over 3.1 million SNPs. Nature **449**, 851–862 (2007)
53. Han, F., Pan, W.: A data-adaptive sum test for disease association with multiple common or rare variants. Hum. Hered. **70**(1), 42–54 (2010)
54. Hansen, M.H., Kooperberg, C.: Spline adaptation in extended linear models (with discussion). Stat. Sci. **17**, 2–51 (2002)
55. He, Q., Lin, D.: A variable selection method for genome-wide association studies. Bioinformatics **27**(1), 1–8 (2011)
56. Hindorff, L.A., Junkins, H.A., Hall, P.N., Mehta, J.P., Manolio, T.A.: A Catalog of Published Genome-Wide Association Studies. www.genome.gov/gwastudies
57. Hirschhorn, J.N., Daly, M.J.: Genome-wide association studies for common diseases and complex traits. Nat. Rev. Genet. **6**(2), 95–108 (2005)
58. Hoggart, C.J., Whittaker, J.C., De Iorio, M., Balding, D.J.: Simultaneous analysis of all SNPs in genome-wide and re-sequencing association studies. PLOS Genet. **4**(7), e1000130 (2008). doi:10.1371/journal.pgen.1000130
59. Hothorn, L.A., Hothorn, T.: Order-restricted scores test for the evaluation of population-based case-control studies when the genetic model is unknown. Biometrical J. **51**(4), 659–669 (2009)
60. Iyengar, S.K., Elston, R.C.: The genetic basis of complex traits: rare variants or "common gene, common disease"? Methods Mol. Biol. **376**, 71–84 (2007)
61. Kang, H.M., Zaitlen, N.A., Wade, C.M., Kirby, A., Heckerman, D., Daly, M.J., Eskin, E.: Efficient control of population structure in model organism association mapping. Genetics **178**(3), 1709–1723 (2008)
62. Kang, H.M., Sul, J.H., Service, S.K., Zaitlen, N.A., Kong, S.Y., Freimer, N.B., Sabatti C., Eskin, E.: Variance component model to account for sample structure in genome-wide association studies. Nat. Genet. **42**(4), 348–354 (2010)
63. Kennedy, G.C., Matsuzaki, H., Dong, S., Liu, W.M., Huang, J., Liu, G., Su, X., Cao, M., Chen, W., Zhang, J., Liu, W., Yang, G., Di, X., Ryder, T., He, Z., Surti, U., Phillips, M.S., Boyce-Jacino, M.T., Fodor, S.P., Jones, K.W.: Large-scale genotyping of complex DNA. Nat. Biotechnol. **21**, 1233–1237 (2003)
64. Kooperberg, C., LeBlanc, M., Obenchain, V.: Risk prediction using genome-wide association studies. Genet. Epidem. **34**, 643–652 (2010)
65. Kooperberg, C., Ruczinski, I.: Identifying interacting SNPs using Monte Carlo logic regression. Genet. Epidemiol. **28**(2), 157–170 (2005)
66. Koren, M., Kimmel, G., Ben-Asher, E., Gal, I., Papa, M.Z., Beckmann, J.S., Lancet, D., Shamir, R., Friedman, E.: ATM haplotypes and breast cancer risk in Jewish high-risk women. Br. J. Cancer. **94**(10), 1537–1543 (2006)
67. Lao, O., et al.: Genome-wide association studies for complex traits: consensus, uncertainty and challenges. Curr. Biol. **18**(16), 1241–1248 (2008)

68. Laurie, C.L., et al.: Quality control and quality assurance in genotypic data for genome-wide association studies. Genet. Epidemiol. **34**, 591–602 (2010)
69. Li, J., Das, K., Fu, G., Li, R., Wu, R.: The Bayesian Lasso for genome-wide association studies. Bioinformatics **27**(4), 516–523 (2010)
70. Li, B., Leal, S.M.: Methods for detecting associations with rare variants for common diseases: application to analysis of sequence data. Am. J. Hum. Genet. **83**(3), 311–321 (2008)
71. Lin, S., Carvalho, B., Cutler, D.J., Arking, D.E., Chakravarti, A., Irizarry, R.A.: Validation and extension of an empirical Bayes method for SNP calling on affymetrix microarrays. Genome Biol. **9**, R63 (2008)
72. Lippert, C., Listgarten, J., Liu, Y., Kadie, C.M., Davidson, R.I., Heckerman, D.: FaST linear mixed models for genome-wide association studies. Nat. Methods **8**(10), 833–835 (2011)
73. Liu, W., Di, X., Yang, G., Matsuzaki, H., Huang, J., Mei, R., Ryder, T.B., Webster, T.A., Dong, S., Liu, G., Jones, K.W., Kennedy, G.C., Kulp, D.: Algorithms for large-scale genotyping microarrays. Bioinformatics **19**, 2397–2403 (2003)
74. Long, J.C.: The genetic structure of admixed populations. Genetics **127**, 417–428 (1991)
75. Lou, X.Y., Chen, G.B., Yan, L., Ma, J.Z., Zhu, J., et al.: A generalized combinatorial approach for detecting gene-by-gene and gene-by-environment interactions with application to nicotine dependence. Am. J. Hum. Genet. **80**, 1125–1137 (2007)
76. Manolio, T.A., et al.: Finding the missing heritability of complex diseases. Nature **461**(7265), 747–753 (2009)
77. Marchini, J., Donnelly, P., Cardon, L.R.: Genome-wide strategies for detecting multiple loci that influence complex diseases. Nat. Genet. **37**(4), 413–417 (2005)
78. Marchini, J., Howie, B.: Genotype imputation for genome-wide association studies. Nat. Rev. Genet. **11**, 499–511 (2010)
79. McCarthy, M.I., Abecasis, G.R., Cardon, L.R., Goldstein, D.B., Little, J., Ioannidis, J.P., Hirschhorn, J.N.: Genome-wide association studies for complex traits: consensus, uncertainty and challenges. Nat. Rev. Genet. **9**(5), 356–369 (2008)
80. McCarthy, M.I., Hirschhorn, J.N.: Genome-wide association studies: potential next steps on a genetic journey. Hum. Mol. Genet. **17**, R156–R165 (2008)
81. McCullagh, P., Nelder, J.A.: Generalized Linear Models, 2nd edn. Chapman and Hall/CRC, Boca Raton (1989)
82. McKeigue, P.M.: Mapping genes underlying ethnic differences in disease risk by linkage disequilibrium in recently admixed populations. Am. J. Hum. Genet. **60**(1), 188 (1997)
83. Meinshausen, N., Bhlmann, P.: Stability selection. JRSSB **72**, 417–448 (2010)
84. Menozzi, P., Piazza, A., Cavalli-Sforza, L.: Synthetic maps of human gene frequencies in Europeans. Science **201**, 786–792 (1978)
85. Miller, D.J., Zhang, Y., Yu, G.: An algorithm for learning maximum entropy probability models of disease risk that efficiently searches and sparingly encodes multilocus genomic interactions. Bioinformatics **25**(19), 2478–2485 (2009)
86. Moore, J.H.: The ubiquitous nature of epistasis in determining susceptibility to common human diseases. Hum. Hered. **56**, 73–82 (2003)
87. Moore, J.H., Gilbert, J.C., Tsai, C.T., Chiang, F.T., Holden, T., Barney, N., White, B.C.: A flexible computational framework for detecting, characterizing, and interpreting statistical patterns of epistasis in genetic studies of human disease susceptibility. J. Theor. Biol. **241**(2), 252–261 (2006)
88. Morgenthaler, S., Thilly, W.G.: A strategy to discover genes that carry multi-allelic or mono-allelic risk for common diseases: a cohort allelic sums test (CAST). Mutat. Res. **615**(1–2), 28–56 (2007)
89. National Center for Biotechnology Information, United States National Library of Medicine. NCBI dbSNP build 144 for human. Summary Page. http://www.ncbi.nlm.nih.gov/projects/SNP/snp_summary.cgi?view+summary=view+summary&build_id=144. Accessed 26 Aug 2015
90. Nelson, M.R., et al.: The population reference sample, POPRES: a resource for population, disease, and pharmacological genetics research. Am. J. Hum. Genet. **83**, 347–358 (2008)

91. Ouwehand, W.H.: The discovery of genes implicated in myocardial infarction. J. Thromb. Haemost. **7**(Suppl 1), 305–307 (2009)
92. Park, T., Casella, G.: The Bayesian Lasso. JASA **103**, 681–686 (2008)
93. Pattin, K.A., White, B.C., Barney, N., et al.: A computationally efficient hypothesis testing method for epistasis analysis using multifactor dimensionality reduction. Genet. Epidemi. **33**(1), 87–94 (2009)
94. Pierce, J.R.: An Introduction to Information Theory: Symbols, Signals, and Noise. Dover, New York (1980)
95. Potkin, S.G., Turner, J.A., Guffanti, G., Lakatos, A., Torri, F., Keator, D.B., Macciardi, F.: Genome-wide strategies for discovering genetic influences on cognition and cognitive disorders: methodological considerations. Cogn. Neuropsychiatry **14**(4/5), 391–418 (2009)
96. Pritchard, J.K., Rosenberg, N.A.: Use of unlinked genetic markers to detect population stratification in association studies. Am. J. Hum. Genet. **65**, 220–228 (1999)
97. Pritchard, J., Stephens, M., Donnelly, P.: Inference of population structure using multilocus genotype data. Genetics **155**(2), 945 (2000)
98. Price, A.L., et al.: Principal components analysis corrects for stratification in genome-wide association studies. Nat. Genet. **38**, 904–909 (2006)
99. Price, A.L., Patterson, N., Yu, F., et al.: A genomewide admixture map for Latino populations. Am. J. Hum. Genet. **80**(6), 1024–1036 (2007)
100. Price, A.L., Tandon, A., Patterson, N., Barnes, K.C., Rafaels, N., Ruczinski, I., Beatty, T.H., Mathias, R., Reich, D., Myers, S.: Sensitive detection of chromosomal segments of distinct ancestry in admixed populations. PLoS Genet. **5**(6), e1000519 (2009)
101. Price, A.L., Zaitlen, N.A., Reich, D., Patterson, N.: New approaches to population stratification in genome-wide association studies. Nat. Rev. Genet. **11**(7), 459–463 (2010)
102. Purcell, S., Neale, B., Todd-Brown, K., et al.: PLINK: a tool set for whole-genome association and population-based linkage analyses. Am. J. Hum. Genet. **81**(3), 559–575 (2007)
103. Rabbee, N., Speed, T.P.: A genotype calling algorithm for affymetrix SNP arrays. Bioinformatics **22**, 7–12 (2006)
104. Redden, D.T., Divers, J., Vaughan, L.K., et al.: Regional admixture mapping and structured association testing: conceptual unification and an extensible general linear model. PLoS Genet. **2**, e137 (2006)
105. Reich, D.E., Goldstein, D.B.: Detecting association in a case-control study while correcting for population stratification. Genet. Epidemiol. **20**, 4–16 (2001)
106. Ritchie, M.E., Carvalho, B.S., Hetrick, K.N., Tavaré, S., Irizarry, R.A.: R/Bioconductor software for Illumina's Infinium whole-genome genotyping BeadChips. Bioinformatics **25**, 2621–2623 (2009)
107. Ritchie, M.D., Hahn, L.W., Roodi, N., Bailey, L.R., Dupont, W.D., Parl, F.F., Moore, J.H.: Multifactor-dimensionality reduction reveals high-order interactions among estrogen-metabolism genes in sporadic breast cancer. Am. J. Hum. Genet. **69**(1), 138–147 (2001)
108. Riveros, C., Vimieiro, R., Holliday, E.G.: Identification of Genome-Wide SNP-SNP and SNP-Clinical Boolean Interactions in Age-Related Macular Degeneration In Epistasis, 217–255. Springer, New York (2015)
109. Robertson, T., Wright, F.T., Dykstra, R.L.: Order Restricted Statistical Inference. Wiley, New York (1988)
110. Nature Genetics Genome-wide association analysis of metabolic traits in a birth cohort from a founder population. **41**(1), 35–46 (2009)
111. Sampson, J.N., Zhao, H.: Genotyping and inflated type I error rate in genome-wide association case/control studies. BMC Bioinform. **10**, 68 (2009)
112. Sasieni, P.D.: From genotypes to genes: doubling the sample size. Biometrics **53**, 1253–1261 (1997)
113. Scheet, P., Stephens, M.: A fast and flexible statistical model for large-scale population genotype data: applications to inferring missing genotypes and haplotypic phase. Am. J. Hum. Genet. **78**, 629–644 (2006)

114. Schwender, H., Ickstadt, K.: Identification of SNP interactions using logic regression. Bio-statistics **9**(1), 187–198 (2008)
115. Schwender, H., Ruczinski, I., Ickstadt, K.: Testing SNPs and sets of SNPs for importance in association studies. Biostatistics (2010). doi:10.1093/biostatistics/kxq042
116. Segura, V., Vilhjalmsson, B.J., Platt, A., Korte, A., Seren, Ü., Long, Q., Nordborg, M.: An efficient multi-locus mixed-model approach for genome-wide association studies in structured populations. Nat. Genet. **44**(7), 825–830 (2012)
117. Setakis, E., Stirnadel, H., Balding, D.J.: Logistic regression protects against population structure in genetic association studies. Genome Res. **16**, 290–296 (2006)
118. Spielman, R.S., McGinnis, R.E., Ewens, W.J.: Transmission test for linkage disequilibrium: the insulin gene region and insulin-dependent diabetes mellitus (IDDM). Am. J. Hum. Genet. **52**(3), 506–516 (1993)
119. Stranger, B.E., Nica, A.C., Forrest, M.S., Dimas, A., Bird, C.P., Beazley, C., Ingle, C.E., Dunning, M., Flicek, P., Montgomery, S., Tavaré, S., Deloukas, P., Dermitzakis, E.T.: Population genomics of human gene expression. Nat. Genet. **39**, 1217–1224 (2007)
120. Szulc, P., Bogdan, M., Frommlet, F., Tang H.: Joint Genotype- and Ancestry-based Genome-wide Association Studies in Admixed Populations. *Working Paper* (2015)
121. Tang, H., Siegmund, D.O., Johnson, N.A., Romieu, I., London, S.J.: Joint testing of genotype and ancestry association in admixed families. Genet. Epidemiol. **34**(8), 783–791 (2010)
122. Tibshirani, R.: Regression shrinkage and selection via the lasso. J. Roy. Stat. Soc. B **58**(1), 267–288 (1996)
123. Via, M., Gignoux, C., Burchard, E.G.: The 1000 genomes project: new opportunities for research and social challenges. Genome Med. **2**, 3 (2010)
124. Wei, Z., Sun, W., Wang, K., Hakonarson, H.: Multiple testing in genome-wide association studies via hidden Markov models. Bioinformatics **25**(21), 2802–2808 (2009)
125. Wolf, B.J., Hill, E.G., Slate, E.H.: Logic forest: an ensemble classifier for discovering logical combinations of binary markers. Bioinformatics **26**(17), 2183–2189 (2010)
126. Wu, T.T., Chen, Y.F., Hastie, T., Sobel, E., Lange, K.: Genome-wide association analysis by lasso penalized logistic regression. Bioinformatics **25**(6), 714–721 (2009)
127. Yang, C., He, Z., Wan, X., Yang, Q., Xue, H., Yu, W.: SNPHarvester: a filtering-based approach for detecting epistatic interactions in genome-wide association studies. Bioinformatics **25**(4), 504–511 (2009)
128. Yang, J., et al.: Common SNPs explain a large proportion of heritability for human height. Nat. Genet. **42**, 565–569 (2010)
129. Yu, J., Pressoir, G., Briggs, W.H., Vroh Bi, I., Yamasaki, M., Doebley, J.F., McMullen, M.D., Gaut, B.S., Nielsen, D.M., Holland, J.B., Kresovich, S., Buckler, E.S.: A unified mixed-model method for association mapping that accounts for multiple levels of relatedness. Nat. Genet. **38**(2), 203–208 (2006)
130. Żak-Szatkowska, M., Bogdan, M.: Modified versions of Bayesian information criterion for sparse generalized linear models. CSDA. In Press, Accepted Manuscript (2012)
131. Zehetmayer, S., Posch, M.: False discovery rate control in two-stage designs. BMC Bioinform. **6**13, 81 (2012). doi:10.1186/1471-2105-13-81
132. Zhang, Y., Liu, J.S.: Bayesian inference of epistatic interactions in case-control studies. Nat. Genet. **39**(9), 1167–1173 (2007)
133. Zhao, J., Chen, Z.: A two-stage penalized logistic regression approach to case-control genome-wide association studies. www.stat.nus.edu.sg/~stachenz/MS091221PR.pdf (2010)
134. Ziegler, A., König, I.R., Thompson, J.R.: Biostatistical aspects of genome-wide association studies. Biometrical J. **50**(1), 8–28 (2008)

Chapter 6
Appendix A: Basic Statistical Distributions

In this section, we recall some probability distributions that play a major role in statistical inference. The material presented here is rather basic, and a more comprehensive treatment can be found in almost any introductory book on probability, for example, in [1]. First, we consider some important continuous random variables. These are variables that can take any value in a given range. Their distribution is described by a density function. Afterwards, we present some important discrete random variables. These variables take integer values. Their distribution is described by a probability mass function.

6.1 Normal Distribution

Suppose the random variable X has a normal distribution with expected value μ and variance σ^2 (or equivalently standard deviation σ). The density function of this distribution is given by the function f, where

$$f(x) = \frac{1}{\sqrt{2\pi}\sigma} \exp\left(-\frac{(x-\mu)^2}{2\sigma^2}\right), \ x \in \mathbf{R}.$$

We write $X \sim N(\mu, \sigma^2)$ (in words X has a normal distribution with the two given parameters). Then the random variable $Z = \frac{X-\mu}{\sigma}$ has a standard normal distribution with $\mu_Z = 0$ and $\sigma_Z = 1$. Its density function ϕ has the form

$$\phi(x) = \frac{1}{\sqrt{2\pi}} \exp\left(-\frac{x^2}{2}\right). \tag{6.1}$$

The cumulative distribution function (cdf) of the standard normal distribution is given by the function Φ, where

© Springer-Verlag London 2016
F. Frommlet et al., *Phenotypes and Genotypes*, Computational Biology 18,
DOI 10.1007/978-1-4471-5310-8_6

$$\Phi(x) = P(Z \le x) = \int_{-\infty}^{x} \phi(x)dx, \tag{6.2}$$

whose value can be calculated routinely by any statistical package, as well as by most modern calculators.

A normal random sample:

A normal random sample of size n is defined to be a sequence $X_1,\dots, X_n$ of independent, identically distributed (i.i.d.) variables from the normal distribution $\mathcal{N}(\mu, \sigma^2)$.

In many cases μ and σ^2 are unknown. To estimate these parameters, one generally uses the statistics

$$\hat{\mu} = \bar{X} = \frac{1}{n}\sum_{i=1}^{n} X_i \text{ and } \hat{\sigma}^2 = s^2 = \frac{1}{n-1}\sum_{i=1}^{n}(X_i - \bar{X})^2. \tag{6.3}$$

These estimates are unbiased, i.e., $E(\bar{X}) = \mu$ and $E(s^2) = \sigma^2$.

Distribution of the sample mean:

Let $X_1, \dots, X_n$ be a normal random sample from $\mathcal{N}(\mu, \sigma^2)$. Then the sample mean $\bar{X}$ has the normal distribution $N\left(\mu, \frac{\sigma^2}{n}\right)$ and the standardized random variable

$$Z := \sqrt{n}\,\frac{\bar{X} - \mu}{\sigma} \tag{6.4}$$

has a standard normal distribution.

The importance of the normal distribution in practical applications is justified by the Central Limit Theorem. In its classical form, this theorem describes the asymptotic distribution of the mean of a general (not necessarily normal) random sample.

Central Limit Theorem

Let $X_1, \dots, X_n$ be a sequence of i.i.d. random variables with expected value μ and standard deviation $\sigma \in (0, \infty)$. Then the cdf of the standardized sample mean converges to the cdf of a standard normal distribution:

$$P\left(\sqrt{n}\,\frac{\bar{X} - \mu}{\sigma} \le x\right) \to \Phi(x) \text{ as } n \to \infty.$$

In intuitive terms, this theorem says that the mean of a large sample (or equivalently the sum of a large number of i.i.d. observations) has approximately a normal distribution, independently of the distribution of the individual observations.

The Central Limit Theorem and its extensions indicate why the normal distribution appears naturally in many contexts of statistical applications. For example, if a trait of interest results from the joint influence of many random components, then this trait can often be well modeled by a normal random variable. It is important to observe that usually such an approximation works quite well, even for relatively small sample sizes. Often, $n \geq 30$ suffices to obtain a good approximation.

However, it is not always true that the normal approximation is applicable. The most common reason for the Central Limit Theorem not to hold is the presence of outliers. As a consequence, in practice one should always carefully check data for unusually large or unusually small observations within the study population. If such observations are present, classical statistical methods relying on the Central Limit Theorem might fail to work properly.

6.2 Important Distributions of Sample Statistics

Apart from the normal distribution, the three most important distributions that arise in the context of sample statistics are the chi-square distribution, Student's t-distribution, and the F-distribution.

6.2.1 Chi-Square Distribution

Let $Z_1, \ldots, Z_k$ be independent standard normal variables. Then the random variable

$$\chi^2 = \sum_{i=1}^{k} Z_i^2$$

has a chi-square distribution with k degrees of freedom, often denoted as χ_k^2. The density function of this distribution is of the form

$$f(x; k) = \frac{x^{(k/2)-1} e^{-x/2}}{2^{k/2} \Gamma(k/2)} \quad \text{for } x \geq 0, \tag{6.5}$$

where the gamma function Γ is defined as

$$\Gamma(t) = \int_0^\infty e^{-x} x^{t-1} \, dx. \tag{6.6}$$

Chi-square distributions are special cases of gamma distributions (see Sect. 6.3).

The expected value of the chi-square distribution with k degrees of freedom is equal to k and its variance is equal to $2k$. From the Central Limit Theorem, for large k, the chi-square distribution can be approximated by a normal distribution with $\mu = k$ and $\sigma = \sqrt{2k}$. The chi-square distribution often occurs in the context of estimating variances:

Distribution of the sample variance:

Let $X_1, \ldots, X_n$ be a normal random sample from $\mathcal{N}(\mu, \sigma^2)$. Let s^2 be the unbiased estimator of σ^2 given by (6.3). Then $\frac{(n-1)s^2}{\sigma^2}$ has a chi-square distribution with $n - 1$ degrees of freedom.

6.2.2 Student's t-Distribution

The density function of Student's t-distribution with ν degrees of freedom is defined as

$$f(x) = \frac{\Gamma\left(\frac{\nu+1}{2}\right)}{\sqrt{\nu\pi}\,\Gamma\left(\frac{\nu}{2}\right)} \left(1 + \frac{x^2}{\nu}\right)^{-\frac{\nu+1}{2}}, \tag{6.7}$$

where Γ is again the gamma function defined in (6.6). Similar to the standard normal distribution, the density of the t-distribution is bell shaped and symmetric around zero, but especially for small ν it has substantially heavier tails than the normal distribution.

If X has a t-distribution with ν degrees of freedom, then $E|X|^\nu = \infty$. In particular, Student's t-distribution with one degree of freedom is the well-known Cauchy distribution, which does not have an expected value. If $\nu \le 2$, then the t-distribution does not have a finite variance, and thus the Central Limit Theorem does not apply. However, for larger values of ν the tails of the t-distribution become less heavy, and as $\nu \to \infty$ the t-distribution converges towards the standard normal distribution.

Consider a normal random sample of size n with unknown μ and σ. According to (6.3), we estimate μ by the sample mean $\bar{X}$, and σ^2 by s^2. Then the standard deviation of $\bar{X}$, which equals $\frac{\sigma}{\sqrt{n}}$, can be estimated by the standard error of the mean SE, where

$$SE = \frac{s}{\sqrt{n}}.$$

This expression can be used to standardize the sample mean when the variance is unknown.

Distribution of the standardized sample mean when the variance is unknown:

For a normal random sample of size n, the statistic

$$t = \frac{\bar{X} - \mu}{SE}$$

has a t-distribution with $df = n - 1$ degrees of freedom.

As a consequence, the t-distribution is often used to test for the equality of means and to build confidence intervals for μ. Because the t-distribution converges to the standard normal distribution as the number of degrees of freedom increases, in many practical applications the t-distribution is actually replaced by the normal distribution when $n \geq 30$.

6.2.3 F-distribution

Let $X_n \sim \chi_n^2$ and $X_d \sim \chi_d^2$ be two independent chi-square random variables with n and d degrees of freedom, respectively. The statistic

$$F = \frac{X_n/n}{X_d/d}$$

is said to have an F-distribution with n degrees of freedom in the numerator and d degrees of freedom in the denominator (or more succinctly, with n and d degrees of freedom).

The density function of the F-distribution with n and d degrees of freedom takes the form

$$f(x; n, d) = \frac{1}{x B(n/2, d/2)} \sqrt{\frac{(nx)^n d^d}{(nx + d)^{n+d}}} \quad \text{for } x \geq 0, \tag{6.8}$$

where B denotes the beta function $B(a, b) = \int_0^1 x^{a-1}(1 - x)^{b-1} \, dx$.

The F-distribution can be used to compare different variances, as in the classical analysis of variance (ANOVA) setting presented in Sect. 8.3.1. It also plays a major role within the context of hypothesis testing in the context of linear models.

6.3   Gamma and Beta Distributions

We have already mentioned that the chi-square distribution is a special case within the family of gamma distributions. The density function of a gamma distribution with shape parameter α and rate parameter β is given by the formula

$$f(x; \alpha, \beta) = \frac{\beta^\alpha}{\Gamma(\alpha)} x^{\alpha-1} e^{-\beta x} \quad \text{for } x \geq 0.$$

When $\alpha = 1$, the gamma distribution reduces to an exponential distribution, discussed below. When $\beta = \frac{1}{2}$ and $\alpha = \frac{k}{2}$, where k is a natural number, we obtain the chi-square distribution with k degrees of freedom.

   The gamma distribution is sometimes used to model the time of failure (or death) when the intensity of failures changes with time. In Bayesian statistics, it is often used as the conjugate prior for the precision (reciprocal of the variance) of the normal distribution.

6.3.1   Exponential Distribution

The density function of the exponential distribution with parameter γ is given by the formula

$$f(x; \gamma) = \gamma \, e^{-\gamma x} \quad \text{for } x \geq 0. \tag{6.9}$$

The expected value of this distribution and its standard deviation are both equal to $1/\gamma$. For $\gamma = 1/2$, the exponential distribution is equal to the chi-square distribution with two degrees of freedom.

   The exponential distribution naturally occurs as the time between events in Poisson processes where events occur at mean rate γ, see also Eq. 6.11 below. Therefore, waiting times are typical examples of exponentially distributed variables. However, in our context such a random variable may be interpreted as the physical distance between two genetic loci where a mutation has occurred. In reliability theory, the exponential distribution is used to model the failure time in situations where the failure intensity γ is constant over time.

6.3.2   Inverse Gamma Distribution

The density function of the inverse gamma distribution is given by the formula

$$f(x; \alpha, \beta) = \frac{\beta^\alpha}{\Gamma(\alpha)} x^{-\alpha-1} e^{-\beta/x} \quad \text{for } x > 0.$$

If X has a gamma distribution with parameters α and β, then X^{-1} has an inverse gamma $(\alpha, \frac{1}{\beta})$ distribution. The inverse gamma distribution often serves in Bayesian statistics as a conjugate prior for the variance of the normal distribution.

6.3.3 Beta Distribution

The beta distribution with parameters α and β has the density function

$$f(x; \alpha, \beta) = \frac{x^{\alpha-1}(1-x)^{\beta-1}}{B(\alpha, \beta)} \quad \text{for } x \in (0, 1),$$

where B is again the beta function (see Sect. 6.2.3). When $\alpha = \beta = 1$, then the beta distribution reduces to the uniform distribution on $[0,1]$, $\mathcal{U}(0, 1)$. Its density is simply the constant function $f(x) = 1$ for $x \in [0, 1]$, and its cdf is $F(x) = x$ for $x \in [0, 1]$. The beta distribution is often used to model unknown proportions. For example, in the Bayesian analysis of high-dimensional data, the beta distribution may be used to model the sparsity parameter.

6.4 Double Exponential Distribution and Extensions

The density function of the double exponential distribution (or Laplace distribution) with scale parameter ξ is given by

$$f(x; \xi) = \frac{\xi}{2} e^{-\xi \, |x|}.$$

The double exponential distribution is symmetric around zero and has variance $2\xi^{-2}$. It is sometimes used in Bayesian statistics to model the magnitude of unknown effects. The Laplace distribution is particularly useful when one expects that the distribution of these effects has tails which are substantially "heavier" than the tails of the normal distribution (i.e., some outliers are observable). A more detailed description is given in Sect. 5.4.2, where we also introduce the normal exponential gamma distribution (NEG), which is slightly less common than the double exponential distribution.

6.4.1 Asymmetric Double Exponential (ADE) Distribution

A particular generalization of the Laplace distribution is the asymmetric double exponential (ADE) distribution with density function defined as

$$f(x; \xi_1, \xi_2) = \begin{cases} \xi_1/2 \ e^{\xi_1 x} & x \le 0 \\ \xi_2/2 \ e^{-\xi_2 x} & x > 0 \end{cases} \quad \xi_1 > 0, \xi_2 > 0.$$

In particular, the ADE distribution can be used to model situations where one believes that strong signals with a positive sign occur more (or possibly less) frequently than strong signals with a negative sign.

6.5 Discrete Distributions

Here, we recall some properties of several well-known discrete distributions. The binomial, Poisson, and negative binomial distributions are among the most widely used discrete distributions. The generalized Poisson and zero-inflated generalized Poisson distributions are perhaps less well known, but they play an important role in modeling specific traits in QTL mapping (see Sect. 4.2.3).

6.5.1 Binomial Distribution

The binomial distribution naturally occurs as the distribution of the number of successes in n independent (Bernoulli) experiments, where the success probability p is the same for each experiment. Let X be the number of successes, then the probability mass function of the binomial distribution $B(n, p)$ is

$$P(X = k) = \binom{n}{k} p^k (1 - p)^{n-k}, \quad k \in \{1, \ldots, n\}. \tag{6.10}$$

The expected value and the variance of $B(n, p)$ are equal to

$$E(X) = np, \quad \sigma_X^2 = np(1 - p).$$

From the Central Limit Theorem, the binomial distribution can be approximated by the normal distribution $\mathcal{N}(np, np(1 - p))$ when n is sufficiently large. As a rule of thumb, this approximation can be applied when $np \geq 5$ and $n(1 - p) \geq 5$.

Note that since a continuous distribution is used to approximate a discrete distribution, one commonly applies a continuity correction. Suppose X has a discrete distribution on the set of integers. In order to approximate $P(X = k)$, where k is an integer, we use the fact that $P(X = k) = P(k - 1/2 < X < k + 1/2)$.

6.5.2 Poisson Distribution

The Poisson distribution is often used to model the number of "rare events" that occur in a certain period. Its probability mass function is given by the formula

$$P(X = k) = \frac{\lambda^k}{k!} \exp(-\lambda), \quad k = 0, 1, 2, \ldots. \tag{6.11}$$

The expected value and the variance of the Poisson distribution are both equal to λ.

If the waiting time for consecutive "rare" events has an exponential distribution with parameter γ, as introduced in Eq. (6.9) of Sect. 6.3.1, then the number of such events in an interval of length T has a Poisson distribution with parameter $\lambda = \gamma T$.

For large n and small p, the Binomial distribution $B(n, p)$ can be approximated by the Poisson distribution with parameter $\lambda = np$. As a rule of thumb, this approximation can be applied whenever $n \geq 20$ and $p \leq 0.05$.

6.5.3 Negative Binomial Distribution

The negative binomial distribution $NB(r, p)$ with parameters r and p describes the distribution of the number of successes that occur in a series of Bernoulli experiments which is performed until the r-th failure is observed. Its probability mass function is given as

$$P(X = k) = \binom{k + r - 1}{k} p^k (1 - p)^r, \quad k = 0, 1, 2, \ldots.$$

For $r = 1$ the negative binomial distribution is equal to the geometric distribution. The expected value and the variance of the negative binomial distribution are equal to

$$E(X) = \frac{rp}{1 - p}, \quad \mathrm{Var}(X) = \frac{rp}{(1 - p)^2}.$$

Thus the mean of the negative binomial distribution can be different from its variance. Due to this flexibility, the negative binomial distribution is often used to replace the Poisson distribution when modeling the number of rare events.

6.5.4 Generalized Poisson Distribution

The generalized Poisson distribution (GP) is another distribution used to model the number of rare events when the mean is different from the variance. Its probability mass function is given as

$$P(X = k; \mu, \varphi) = \frac{\mu(\mu + (\varphi - 1)k)^{k-1}}{k!} \varphi^{-k} e^{-\frac{1}{\varphi}(\mu + (\varphi - 1)k)}, \quad k = 0, 1, 2, \ldots,$$

$$(6.12)$$

where μ and φ are larger than 0. For $Y \sim GP(\mu, \varphi)$, we have $E(Y) = \mu$ and $\mathrm{Var}(Y) = \varphi^2 \mu$. This allows us to model over- and underdispersion (which correspond to the variance of the number of rare events being larger and smaller than the expected number, respectively).

6.5.5   Zero-Inflated Generalized Poisson Distribution

In many examples of rare events, the probability that the number of such events is equal to zero is substantially larger than the probability given by the Generalized Poisson Distribution. To model such phenomena one can use the Zero-Inflated Generalized Poisson Distribution (ZIGP), which is defined as a mixture of a distribution concentrated at 0, denoted as δ_0, and the generalized Poisson distribution, GP, with probability mass function

$$ZIGP(\mu, \varphi, \omega) = \omega\delta_0 + (1 - \omega)GP(\mu, \varphi). \tag{6.13}$$

Reference

1. Ross, S.: A First Course in Probability, 8th edn. Parson (2010)

Chapter 7
Appendix B: Basic Methods of Estimation

In this chapter, we will briefly discuss some basic notions and methods of classical statistical estimation. For an in-depth treatment we recommend, for example, [1]. We will start with a discussion of some basic properties of statistical estimators, such as unbiasedness and efficiency. Then we will discuss two basic methods of statistical estimation in more detail—maximum likelihood estimation and the method of moments.

7.1 Basic Properties of Statistical Estimators

7.1.1 Statistical Bias

Assume that the elements of the sample $y = (y_1, \ldots, y_n)$ are realizations of n independent random variables $Y_1, \ldots, Y_n$, all with a common density function $f(x; \theta)$. The vector containing the K parameters of the distribution, $\theta = (\theta_1, \theta_2, \ldots, \theta_K)$, takes values from Ω_θ, where Ω_θ is some open subset of $\mathbb{R}^K$. Let us denote a function of the sample y which can be used to estimate θ_i by $\hat{\theta}_i$. Similarly, an estimator of the vector θ will be denoted by $\hat{\theta}$.

Bias of a statistical estimator

Let us consider a single parameter $\theta \in \Omega_\theta \subset R$ and its estimator $\hat{\theta}$. The bias $B_\theta(\hat{\theta})$ of the estimator $\hat{\theta}$ is defined as the difference between its expected value and the true value of the parameter, i.e., in mathematical terms

$$B_\theta(\hat{\theta}) = E_\theta(\hat{\theta}) - \theta.$$

The statistic $\hat{\theta}$ is an unbiased estimator of θ if for every $\theta \in \Omega_\theta$, $B_\theta(\hat{\theta}) = 0$.

© Springer-Verlag London 2016
F. Frommlet et al., *Phenotypes and Genotypes*, Computational Biology 18,
DOI 10.1007/978-1-4471-5310-8_7

7.1.2 Mean Square Error

One of the basic measures of the performance of statistical estimators is the mean square error, which provides information on the average l_2 (square) distance between the estimator and the parameter.

The mean square error
Let us consider a parameter $\theta \in \Omega_\theta \subset R$ and its estimator $\hat{\theta}$. The mean square error $MSE_\theta(\hat{\theta})$ of the estimator $\hat{\theta}$ is defined as

$$MSE_\theta(\hat{\theta}) = E_\theta(\hat{\theta} - \theta)^2 = B_\theta^2(\hat{\theta}) + Var_\theta(\hat{\theta}).$$

It is clear from the above definition that, in terms of MSE, good estimators should have a small absolute bias and a low variance.

7.1.3 Efficiency of Estimators

In this section, we will discuss the notion of efficient estimators. For this purpose, we will first introduce the Fisher information matrix.

Fisher Information Matrix
Consider the family of distributions $f(x; \theta)$ with $\theta \in \Omega_\theta \subset R^K$, such that $f(x; \theta)$ is twice differentiable as a function of θ. The corresponding Fisher information matrix is given by

$$I(\theta) = E\left\{\frac{\partial \log f(x, \theta)}{\partial \theta_i}, \frac{\partial \log f(x, \theta)}{\partial \theta_j}\right\}_{1 \leq i \leq K, 1 \leq j \leq K},$$

where θ_i is the ith parameter of the distribution (the ith component of the vector θ).

The Fisher information matrix plays a major role in the Rao–Cramér inequality, specified below, which provides a lower bound on the variance of the estimators for a regular family of distributions $f(x; \theta)$. The regularity conditions, e.g., [1], require that the support of $f(x, \theta)$ (in intuitive terms, the set of values a random variable can take) does not depend on θ and impose mild smoothness conditions on $f(x; \theta)$ as a function of θ.

Rao-Cramér Inequality.
Let $Y_1, \ldots, Y_n$ be independent random variables with a common density function $f(x; \theta)$, with $\theta \in \Omega_\theta \subset R$. Moreover, assume that $f(x; \theta)$ satisfies classical regularity conditions (e.g., [1]). Let $\hat{\theta}$ be an unbiased estimator of θ. Then

$$Var(\hat{\theta}) \geq \frac{1}{nI(\theta)}.$$

Efficient Estimator.
An unbiased estimator of θ, $\hat{\theta}$, is called efficient if for every $\theta \in \Omega_\theta$,

$$Var(\hat{\theta}) = \frac{1}{nI(\theta)}.$$

Important examples of efficient estimators include the sample mean as the estimator of the expected value for the normal, exponential or Poisson distributions.

7.1.4 Method of Moments

The method of moments is one of the most popular ways of deriving statistical estimators. Let $\theta = (\theta_1, \ldots, \theta_K)$ be the vector of parameters characterizing the distribution of the random variable Y. Then clearly, the vector of the K first moments of Y can be expressed as some vector function of θ, as follows:

$$(EY, EY^2, \ldots, EY^K) = g(\theta_1, \ldots, \theta_K).$$

In the case when the function g is invertible, it follows that

$$(\theta_1, \ldots, \theta_K) = g^{-1}(EY, EY^2, \ldots, EY^K). \tag{7.1}$$

Using the method of moments, the unknown moments of the distribution generating the data in Eq. (7.1) are replaced by the sample moments, based on the sample $y_1, \ldots, y_n$. Thus, the appropriate estimator of θ is given by

$$\hat{\theta} = (\hat{\theta}_1, \ldots, \hat{\theta}_K) = g^{-1}(\frac{1}{n}\sum_{i=1}^{n} y_i, \ldots, \frac{1}{n}\sum_{i=1}^{n} y_i^K).$$

Using the method of moments, it is not necessary that the first K moments are used. In some applications it is better to use other moments. Also, the ith standard moment is often replaced by the corresponding central moment, $E(Y - EY)^i$.

The method of moments is often easy to use. Also the distributions of the resulting estimators are usually easy to derive based on the central limit theorem and the so-called delta method (e.g., [1]). However, it often happens that these estimators are not efficient.

7.1.5 Maximum Likelihood Estimation

Assume that the elements of the sample $y = (y_1, \ldots, y_n)$ are realizations of n independent random variables $Y_1, \ldots, Y_n$, all with a common density function $f(x; \theta)$ with $\theta \in \Omega_\theta \subset R^K$.

The likelihood function:
For a random sample of size n, the likelihood function $\mathscr{L}$ is defined by

$$\mathscr{L}(\theta; y_1, \ldots, y_n) := \prod_{i=1}^{n} f(y_i; \theta). \tag{7.2}$$

The maximum likelihood estimator The maximum likelihood estimator $\hat{\theta}$ of the parameter $\theta = (\theta_1, \ldots, \theta_K)$ is the point at which the likelihood function attains its maximum

$$\hat{\theta} = argmax_{\theta \in \Omega_\theta} \mathscr{L}(\theta; y_1, \ldots, y_n). \tag{7.3}$$

If the density function $f(x; \theta)$ satisfies classical regularity conditions, then the asymptotic distribution of the maximum likelihood estimator converges to the multivariate normal distribution defined below

$$\sqrt{n}(\hat{\theta} - \theta) \to \mathscr{N}(\mathbf{0}, I^{-1}(\theta)),$$

where $\mathbf{0}$ denotes a vector composed of K zeroes. This means that for large sample sizes maximum likelihood estimators are approximately unbiased and efficient.

7.2   Estimates of Basic Statistical Parameters

7.2.1   Mean and Variance

Let the sample $y = (y_1, \ldots, y_n)$ consist of the realizations of n independent random variables $Y_1, \ldots, Y_n$ from the same distribution with expected value equal to μ and variance equal to σ^2. The unbiased estimates of these parameters are given by the following formulas:

$$\hat{\mu} = \bar{y} = \frac{\sum_{i=1}^{n} y_i}{n}$$

and

$$\hat{\sigma}^2 = s^2 = \frac{\sum_{i=1}^{n} (y_i - \bar{y})^2}{n - 1}.$$

7.2.2   Pearson Correlation Coefficient

Let X and Y be two random variables. The population correlation coefficient is defined as

$$\rho = \frac{Cov(X, Y)}{\sigma_x \sigma_y} = \frac{E(XY) - \mu_x \mu_y}{\sigma_x \sigma_y},$$

where μ_x and μ_y denote the expected values of X and Y, while σ_x and σ_y denote the corresponding standard deviations.

Any correlation coefficient always takes values from the interval $[-1, 1]$. If $\rho = 0$, then X and Y are not correlated. Positive values of the correlation coefficient arise when large values of Y typically go together with large values of X, while negative values of the correlation coefficient suggest that large values of Y typically go with small values of X. In the case of the Pearson correlation coefficient, when $\rho = 1$ we observe a perfectly linear relationship between Y and X, whose slope is positive, while $\rho = -1$ corresponds to a perfectly linear relationship with a negative slope.

The correlation coefficient can be estimated based on n pairs of observations $(x_1, y_1), \ldots, (x_n, y_n)$, where the measurements x_i and y_i are both made on the ith individual. The Pearson correlation coefficient r estimates ρ based on the following formula:

$$r = \frac{1}{n - 1} \sum_{i=1}^{n} \left(\frac{x_i - \bar{x}}{s_x} \right) \left(\frac{y_i - \bar{y}}{s_y} \right). \tag{7.4}$$

Reference

1. Hogg, R.V., McKean, J. W. and Craig, A.T.: Introduction to Mathematical Statistics, 6th edn. Pearson Education Inc. (2005)

Chapter 8
Appendix C: Principles of Statistical Testing

Here, we will briefly recap the basic principles of statistical testing. For an in depth treatment we recommend, for example, [3]. We will start with the classical examples of the Z-test and t-test to assess statistical hypotheses concerned with sample means. We then describe the classical approaches to ANOVA and multiple regression, before we show how these procedures can be described using the unifying framework of general linear models (GLMs). After briefly discussing generalized linear models, we will cover the nonparametric counterparts of the t-test, ANOVA and regression. The chapter will conclude with the chi-square test and Fisher's exact test as primary examples of statistical tests adapted to the analysis of qualitative variables.

8.1 Basic Ideas of Statistical Testing: The Z-test

Consider a sequence of trait values $X_1, \ldots, X_n$ from a normal population $\mathcal{N}(\mu, \sigma^2)$, where σ^2 is known. One might be interested in testing whether the population average μ is greater than a certain fixed value μ_0, or not. This question can be formalized as the following one-sided test:

One-sided (right tail) test for a mean:

$$H_0 : \mu \leq \mu_0 \text{ against } H_A : \mu > \mu_0, \tag{8.1}$$

where H_0 denotes the null hypothesis and H_A denotes the alternative hypothesis. Using a statistical test, we only reject the null hypothesis H_0 in favor of the alternative hypothesis H_A when there is strong evidence that the alternative is correct.

The natural test statistic for testing any hypothesis about μ is the sample mean $\bar{X}$, or its standardized version Z defined in Eq. (6.4). The test statistic is used to decide

© Springer-Verlag London 2016
F. Frommlet et al., *Phenotypes and Genotypes*, Computational Biology 18,
DOI 10.1007/978-1-4471-5310-8_8

whether to accept or reject the null hypothesis. To this end, the acceptance and the rejection regions need to be defined.

When making a decision based on a statistical test, two kinds of error are possible. It might be the case that H_0 is correct but one accepts H_A. Such a detection of a false positive is called a **type I error**. On the other hand, H_A might be correct, but based on the data, one does not reject H_0. Then one has missed a true signal, which is called a **type II error**. The **power** of a test is defined to be the probability of rejecting H_0 when H_A is correct. Thus the power is simply given by $1 - t_2$, where t_2 is the probability of committing a type II error.

The classical approach to statistical testing is to control the type I error rate at a prespecified level α, the so-called **significance level**. Based on this, a rejection region $B(\alpha)$ is chosen such that $P(\bar{X} \in B(\alpha)|H_0) = \alpha$ (i.e., the probability of belonging to the rejection region given that the null hypothesis is true equals α). Usually, there are many possible choices of the rejection region which control the type I error at a given level. Ideally, one would like to choose this region in such a way that the type II error of the testing procedure is minimized, or equivalently that the power of the test is maximized. According to the Neyman–Pearson theory [6], in many situations the most efficient rejection region is simply an interval.

In the case of the above hypothesis test (8.1), the optimal rejection region is one sided, i.e., one should reject the null hypothesis when $\bar{X}$ is "significantly" larger than μ, which corresponds to large positive values of Z. The boundary of this one-sided rejection interval is called the **critical value**. To derive this critical value, remember from (6.4) that, under the null hypothesis, the test statistic Z has a standard normal distribution $\mathcal{N}(0, 1)$. Thus, to control the type I error at the level α, one should use the following test strategy:

One-sided (right tail) z-test:

Reject the null hypothesis if and only if

$$Z > z_\alpha, \quad \text{where } z_\alpha \text{ is such that } \Phi(z_\alpha) = 1 - \alpha.$$

An important concept of statistical testing is the so-called **p-value**, which is defined to be the smallest significance level under which H_0 is rejected given $Z = z$. Perhaps more intuitively, the p-value can be interpreted as the probability, under the null hypothesis, of observing a value which is equal to or even more extreme than the realization of the test statistic resulting from the data. In the case of such a one-sided Z-test, the p-value is calculated as

$$\text{p-value} = 1 - \Phi(z),$$

where z is the observed realization of the Z-test statistic. This is the appropriate probability of being in the right tail of the distribution. In any statistical test, the p-value can be used to make one's conclusion. The null hypothesis is rejected if

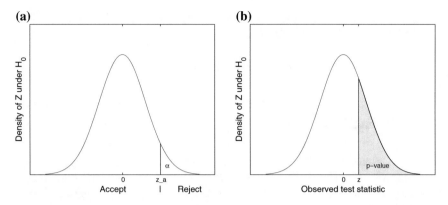

Fig. 8.1 The rejection region and the p-value for a one-sided Z-test. Plot **a** shows the acceptance and rejection regions, where the critical value z_α is chosen such that the probability of a type I error is α. In plot **b** the area of the *yellow* region under the density *curve* is the p-value corresponding to the realization of the test statistic z. Note that the realization of the test statistic is smaller than the critical value and, on the other hand, the p-value is larger than α. Both decision criteria $z < z_\alpha$ and $p > \alpha$ are equivalent and suggest that H_0 should not be rejected

and only if the p-value is less than or equal to the significance level α. Figure 8.1 gives a graphical illustration of the rejection region and the p-value for the one-sided hypothesis test (8.1) based on the Z-test.

Let us now briefly discuss what happens if the hypotheses are reversed, i.e., when we consider the left tail test

$$H_0 : \mu \geq \mu_0 \quad \text{against} \quad H_A : \mu < \mu_0. \tag{8.2}$$

Using the same argument as above, it is easy to check that given $Z = z$ the null hypothesis should be rejected when $z \leq -z_\alpha$. Furthermore, the p-value for the hypothesis test (8.2) is calculated according the formula p-value $= \Phi(z)$, i.e., the appropriate probability of being in the left hand tail of the distribution.

Another important problem is the test for a point null hypothesis regarding a mean against a two-sided alternative:

Point null hypothesis, two-sided alternative:

$$H_0 : \mu = \mu_0 \quad \text{against} \quad H_A : \mu \neq \mu_0. \tag{8.3}$$

In this case, the null hypothesis should be rejected when $|z| \geq z_{\alpha/2}$. This corresponds to a symmetric rejection region, where both large and small realizations of the test statistic Z suggest that the alternative should be chosen. Since the rejection region is symmetric, the critical values are $\pm z_{\alpha/2}$ and H_0 should be accepted when

$z \in (-z_{\alpha/2}, z_{\alpha/2})$. Furthermore, the p-value for the two-sided testing problem (8.3) is calculated according to the formula

$$\text{p-value} = 2(1 - \boldsymbol{\Phi}(|z|)).$$

8.2   The Family of t-tests

T-tests are often the statistical method of choice when one is interested in differences between population means. We recall several classical situations and briefly comment on the extent to which t-tests are still applicable when the usual underlying normality assumptions on the data are not strictly fulfilled.

8.2.1   One Sample t-Test

Let us first consider again the problem of testing for the mean of a normal population $\mathcal{N}(\mu, \sigma^2)$, but in contrast to Sect. 8.1 we now deal with the more common case that the variance is unknown. This results in the familiar t-test.

One sample t-test:

Let $X_i \sim \mathcal{N}(\mu, \sigma^2)$ be i.i.d. for $i = 1, \ldots, n$, where σ is unknown.

To test $H_0 : \mu = \mu_0$, we use

$$T = \frac{\sqrt{n}(\bar{X} - \mu_0)}{s}, \qquad s^2 = \frac{1}{n-1} \sum_{i=1}^{n} (X_i - \bar{X})^2.$$

Under H_0, the test statistic T has a Student t-distribution with $n - 1$ degrees of freedom, $T \sim t_{n-1}$.

The critical regions and p-values for the test statistic T are calculated as in the case of Z-tests, but with the cumulative distribution function of the standard normal distribution $\boldsymbol{\Phi}$ replaced by the cumulative distribution function of the t-distribution. The appropriate values of the distribution function can be calculated in any statistical package.

8.2.2 Two Sample t-Test

In practice, comparison of the means of two different groups is even more relevant. For example, in the case of a backcross design in QTL mapping, a genetic marker can have either genotype AA or genotype Aa (see Sect. 2.2.1.1). A rather simplistic analysis is to test whether the mean of the quantitative trait differs according to the genotype.

The pooled t-test assumes that the variance of the trait is the same in both groups.

Pooled t-test:

Let $X_i \sim \mathcal{N}(\mu_1, \sigma^2)$ be i.i.d. for $i = 1, \ldots, n_1$ and $Y_i \sim \mathcal{N}(\mu_2, \sigma^2)$ be i.i.d. for $i = 1, \ldots, n_2$, where σ is unknown, but assumed to be equal in both groups. To test $H_0 : \mu_1 = \mu_2$, we use

$$T = \frac{(\bar{X} - \bar{Y})}{s_p}, \quad s_p^2 = \left(\frac{1}{n_1} + \frac{1}{n_2}\right) \frac{\sum_{i=1}^{n_1}(X_i - \bar{X})^2 + \sum_{i=1}^{n_2}(Y_i - \bar{Y})^2}{n_1 + n_2 - 2}. \quad (8.4)$$

The critical values and p-values are calculated using the fact that under H_0, $T \sim t_{n_1+n_2-2}$.

If the assumption of equal variances in both groups is violated, then the pooled estimator of variance s_p^2 is replaced by

$$s^2 = \frac{s_1^2}{n_1} + \frac{s_2^2}{n_2},$$

where $s_1^2 = \frac{1}{n_1-1}\sum_{i=1}^{n_1}(X_i - \bar{X})^2$ and $s_2^2 = \frac{1}{n_2-1}\sum_{i=1}^{n_2}(Y_i - \bar{Y})^2$ are the unbiased estimators of the variances in both groups.

It turns out that in this situation the t-test statistic no longer has an exact t-distribution. However, it can be well approximated by a t-distribution with the number of degrees of freedom given by the Welch–Satterthwaite equation,

$$df = \frac{(SE_1^2 + SE_2^2)}{\frac{SE_1^4}{n_1-1} + \frac{SE_2^4}{n_2-1}},$$

where $SE_i = \frac{s_i}{\sqrt{n_i}}$ for $i \in \{1, 2\}$.

8.2.3 Paired t-Test

In some situations, two samples may not be independent. Such a situation occurs, for example, when the goal is to estimate the influence of some form of treatment on blood pressure, and we measure blood pressure in a group of patients before and after treatment. Then the two measurements X_i and Y_i on the ith patient are not independent of each other, but form a "pair". Let us denote by $X_1, \ldots, X_n$ the measurements in the first sample (e.g., observed before treatment) and by $Y_1, \ldots, Y_n$ the measurements in the second sample (e.g., observed after treatment). We want to test whether the average blood pressure after treatment, μ_2, is lower than the average blood pressure before treatment, μ_1. This leads to the following hypothesis test:

$$H_0 : \mu_1 = \mu_2 \quad \text{against} \quad H_A : \mu_1 > \mu_2. \tag{8.5}$$

Note that the hypothesis that we want to prove is specified as the alternative. This is because we should be quite sure that H_A holds before rejecting H_0 (which results from the fact that the probability of a type I error is defined to be small). On the other hand, in the case when H_0 is accepted, we are still not sure that H_0 actually holds. Normally, statistical testing will not indicate any significant results when the average effect size of the treatment, $\mu_1 - \mu_2$, is very small. In general, large sample sizes are necessary to detect small effects via statistical testing. Note also that H_0 must include some form of equality for us to be able to derive the distribution of the test statistic under the null hypothesis.

Since the two samples are not independent, we cannot use the standard two sample t-test. However, it turns out that we have a very simple alternative. Let us define $U_i = Y_i - X_i$ and let μ_U denote the population mean of U. Then the hypothesis (8.5) can be rewritten as

$$H_0 : \mu_U = 0 \quad \text{against} \quad H_A : \mu_U < 0, \tag{8.6}$$

which can be tested using a standard one sample t-test, based on the sample $U_1, \ldots, U_n$.

8.2.4 Robustness of t-Tests

T-tests are usually derived under the assumption that the data come from normal populations. However, due to the central limit theorem, for moderate and large sample sizes t-tests will typically have type I error rates close to the nominal significance level, even if the data do not have a normal distribution. One important indication that a t-test is inappropriate is the presence of outliers, which suggests that the distribution generating the data has "heavy tails" and, as we have discussed in Sect. 6.1, in this case the central limit theorem does not hold. In such a situation, the nonparametric alternatives discussed in Sect. 8.7 should be used.

8.3 Classical Approach to ANOVA and Regression

8.3.1 One-Way Analysis of Variance

Comparing the mean of more than two groups leads to the method called analysis of variance (ANOVA). As an example, consider genome-wide association studies where a genetic marker can have genotypes AA, Aa, or aa (see Sect. 5.3.1). To analyze whether there is an association between a single marker and a quantitative trait, one might compare the mean values of traits in 3 groups. The simplest model assumes that the variance within each group is the same.

In the above example, it is assumed that there is only one factor which influences the trait (the marker genotype). The corresponding analysis is commonly referred to as one factor ANOVA or one-way ANOVA. The classical statistical approach is based on a decomposition of the residual sum of squares, which is illustrated by Table 8.1. Assume that there are k groups, and in group $i, i \in \{1, \ldots, k\}$, there are n_i observations. Let $N = \sum_{i=1}^{k} n_i$ be the total number of observations. Denote by y_{ij} the jth observation from group i. Furthermore, let $\bar{y}_{i.} = \frac{1}{n_i} \sum_{j=1}^{n_i} y_{ij}$ be the mean in the ith group. Finally, the mean over all observations is denoted by $\bar{y} = \frac{1}{N} \sum_{i=1}^{k} \sum_{j=1}^{n_i} y_{ij}$.

We can thus define an (idealized) model in which the value of an observation depends entirely on the factor, i.e., the value of an observation in group i is equal to $\bar{y}_{i.}$. We can thus define the following three sums of squares:

$$MSS := \sum_{i=1}^{k} n_i(\bar{y}_{i.} - \bar{y})^2, \quad \text{Sum of squares from the model}$$

$$RSS := \sum_{i=1}^{k} \sum_{j=1}^{n_i} (y_{ij} - \bar{y}_{i.})^2, \quad \text{Residual sum of squares}$$

$$TSS := \sum_{i=1}^{k} \sum_{j=1}^{n_i} (y_{ij} - \bar{y})^2, \quad \text{Total sum of squares}$$

Table 8.1 Classical one-way ANOVA table

	SS	df	MS	F
Model	MSS	$k - 1$	$MSM = \frac{MSS}{k-1}$	$F = \frac{MSM}{MSR}$
Residual	RSS	$N - k$	$MSR = \frac{RSS}{N-k}$	
Total	$TSS = MSS + RSS$	$N - 1$		

The classical one-way ANOVA procedure is described below:

One factor ANOVA:

Let $Y_{ij} \sim \mathcal{N}(\mu_i, \sigma^2)$ for $i = 1, \ldots, k, j = 1, \ldots, n_i$, where σ is unknown, but assumed to be equal for all k groups. Using the notation for the sum of squares introduced above, the statistic used to test $H_0 : \mu_1 = \mu_2 = \cdots = \mu_k$ is

$$F = \frac{MSS/(k-1)}{RSS/(N-k)} = \frac{MSM}{MSR}. \tag{8.7}$$

Under H_0, the test statistic is F-distributed with $k - 1$ and $N - k$ degrees of freedom, $F \sim F_{k-1, N-k}$.

The principle of one-way ANOVA is to compare the variation in the data between groups (MSS) with the variation in the data within groups (RSS). The decomposition $TSS = MSS + RSS$, which allows us to determine the statistical properties of the test statistic F, is of crucial importance.

An important measure of the influence of the factor in one-way ANOVA (and of the factors in any General Linear Model) is the coefficient of determination

$$R^2 = \frac{MSS}{TSS}, \tag{8.8}$$

which specifies what proportion of the total variance of Y can be explained by the factor (or factors). Further description of one-way analysis of variance can be found in practically any standard statistics textbook. We would recommend, for example, [8].

8.3.2   Two-Way ANOVA. Interactions

Here, we discuss the ANOVA procedure which is concerned with determining the relationship between two qualitative variables (factors) and a continuous dependent variable. Two-way ANOVA allows us to model both the main (individual) effects of the two factors and the interaction between them. An example from genetics would be a trait that is influenced by two different genes. In this context, the interaction effect is commonly referred to as epistasis.

To explain the basics of the two-way ANOVA procedure, we start with an example, a model with two markers based on the backcross design for QTL mapping. As in Sect. 2.2.1.1, we will code the marker genotypes using two dummy variables X_1 and X_2. Let $X_1 = 0$ when an individual is homozygous at the first locus, otherwise $X_1 = 1$. We apply the same definition with respect to the second locus to define X_2.

There are four possible combinations of genotypes $(0, 0)$, $(0, 1)$, $(1, 0)$, $(1, 1)$. These combinations are referred to as cells according to the terminology used for frequency tables. We use $\mu_{ij} = E(Y|X_1 = i, X_2 = j)$, $i, j \in \{0, 1\}$, to denote the expected value of the continuous trait Y given the values of the marker genotypes. To investigate whether the two markers have any influence on the trait, one could apply a one-way ANOVA procedure to test the null hypothesis

$$H_0 : \mu_{00} = \mu_{01} = \mu_{10} = \mu_{11}.$$

Under H_0 all the cell means are identical, and it follows that none of the markers are associated with the trait.

In a general setting, the first factor, A, might have I levels and the second factor, B, might have J levels. So there are $I \cdot J$ cells, and one might accordingly consider a one-way ANOVA procedure with $I \cdot J$ groups to test whether the two factors have any influence on some continuous variable. However, the two-way ANOVA procedure offers a more refined analysis, which allows us to investigate the influence of each factor and the interaction effect separately. This procedure is best explained in terms of the factor effects model.

Two-way ANOVA, factor effects model:

For levels $i = 1, \ldots, I$ and $j = 1, \ldots, J$, let

$$\mu_{ij} = \mu + \alpha_i + \beta_j + \delta_{ij},$$

with the constraints $\sum_{i=1}^{I} \alpha_i = \sum_{j=1}^{J} \beta_j = 0$ on the main effects and $\sum_{i=1}^{I} \delta_{ij} = 0$ for each j and $\sum_{j=1}^{J} \delta_{ij} = 0$ for each i on the interaction effects.

Using this representation, $\mu = \frac{\sum_{ij} \mu_{ij}}{IJ}$ can be interpreted as the "overall" mean of the trait values. The main effect of factor A at level i is equal to $\alpha_i = \frac{\sum_j \mu_{ij}}{J} - \mu$ and can be interpreted as the difference between the mean value of the trait at the ith level of A and the overall mean. Similarly, $\beta_j = \frac{\sum_i \mu_{ij}}{I} - \mu$. Finally, the interaction coefficients δ_{ij} describe the joint influence of the two factors on the trait which cannot be explained by the sum of their main effects. The constraints on the main effects α_i, β_j and the interaction effect δ_{ij} specified above are necessary, because otherwise these parameters would not be uniquely defined and therefore not estimable.

Let us return briefly to our example based on the backcross design, where $I = J = 2$. Table 8.2 gives the values of μ_{ij} for two different traits. In the case of the first trait, the two markers do not interact. It is easy to calculate that in this example $\mu = 75$, $\alpha_1 = -\alpha_0 = 10$, $\beta_1 = -\beta_0 = 15$, and $\delta_{00} = \cdots = \delta_{11} = 0$. The

Table 8.2 Two examples of a two-way ANOVA model

X_2	X_1 0	1		X_2	X_1 0	1
0	50	80		0	50	80
1	70	100		1	30	100

The tables give the values of the cell means μ_{ij}. The first example does not consider an interaction effect, the second example does

lack of any interaction is clearly observed by noticing that the influence of X_1 does not depend on the level of X_2. Specifically, the average value of the trait increases by 30 when the value of X_1 changes from 0 to 1, and this increase does not depend on the value of X_2. Similarly, the average value of the trait increases by 20 when X_2 changes from 0 to 1, and again this increase does not depend on the genotype of the first marker.

The second example in Table 8.2 illustrates a situation where the two markers interact. Here, $\mu = 65$, $\alpha_1 = -\alpha_0 = 25$, $\beta_1 = -\beta_0 = 0$, and $\delta_{00} = \delta_{11} = -\delta_{01} = -\delta_{10} = 10$. Therefore, in this example X_2 does not have a main effect, but influences the trait only by interacting with X_1. Note that the influence of X_1 on the trait is much stronger if $X_2 = 1$ or, equivalently, if an individual is heterozygous at the second marker.

The two-way ANOVA procedure carries out statistical tests of the following three hypotheses:

$$\text{Main effect Factor } A: \quad H_{0A} : \alpha_1 = \cdots = \alpha_I = 0$$
$$\text{Main effect Factor } B: \quad H_{0B} : \beta_1 = \cdots = \beta_J = 0$$
$$\text{Interaction effect:} \quad H_{0AB} : \delta_{11} = \cdots = \delta_{IJ} = 0.$$

The crucial ingredient in this approach is that the sum of squares from the model can be broken into the sum of three components

$$MSS = SSA + SSB + SSAB,$$

where SSA, SSB, and $SSAB$ denote the sums of squares explained by the factors A, B and their interaction, respectively. The details of this analysis are presented, e.g., [8], and one can find an implementation of the two-way ANOVA procedure in any statistical software package. The layout of the classical two-way ANOVA table based on a total number of N observations is presented in Table 8.3.

The test statistic F_A is used to test H_{0A}, the hypothesis that factor A has no influence. Under the null hypothesis H_{0A}, the corresponding test statistic has an F-distribution, $F_A \sim F_{I-1,N-IJ}$. Similarly, F_B is used to test H_{0B}, i.e., whether factor B has no influence, and F_{AB} is used to test the hypothesis of no interaction, H_{0AB}. Under the null hypothesis H_{0B}, $F_B \sim F_{J-1,N-IJ}$ and under the null hypothesis H_{0AB},

Table 8.3 Classical two-way ANOVA table

	SS	df	MS	F
A	SSA	$I-1$	$MSA = \frac{SSA}{I-1}$	$F_A = \frac{MSA}{MSE}$
B	SSB	$J-1$	$MSB = \frac{SSB}{J-1}$	$F_B = \frac{MSB}{MSE}$
AB	$SSAB$	$(I-1)(J-1)$	$MSAB = \frac{SSAB}{(I-1)(J-1)}$	$F_{AB} = \frac{MSAB}{MSE}$
Residual	RSS	$N-IJ$	$MSE = \frac{RSS}{N-IJ}$	
Total	$TSS = MSS + RSS$	$N-1$		

$F_{AB} \sim F_{(I-1)(J-1),N-IJ}$. A null hypothesis is rejected if the corresponding F-test statistic is larger than the one-sided critical value obtained from the F-distribution.

8.3.3 Two-Way ANOVA with No Interactions

If one expects that there are no interactions, or if one accepted the hypothesis H_{0AB} of no interaction, then one may consider the following factor effects model:

$$\mu_{ij} = \mu + \alpha_i + \beta_j,$$

with the constraints $\sum_i \alpha_i = \sum_j \beta_j = 0$. The reduced two-way ANOVA table presented in Table 8.4 allows us to test the following two hypotheses:

$$\text{Main effect Factor } A: \quad H_{0A} : \alpha_1 = \cdots = \alpha_I = 0$$
$$\text{Main effect Factor } B: \quad H_{0B} : \beta_1 = \cdots = \beta_J = 0.$$

The advantage of removing the possibility of interaction is that we can make use of more degrees of freedom to estimate the variance of the residual error. This can help to increase the power of detecting significant main effects.

Table 8.4 Two-way ANOVA table

	SS	df	MS	F
A	SSA	$I-1$	$MSA = \frac{SSA}{I-1}$	$F_A = \frac{MSA}{MSE}$
B	SSB	$J-1$	$MSB = \frac{SSB}{J-1}$	$F_B = \frac{MSB}{MSE}$
Residual	RSS	$N-I-J+1$	$MSE = \frac{RSS}{N-I-J+1}$	
Total	$TSS = MSS + RSS$	$N-1$		

Two main effects, no interactions vs one main effect and an interaction

8.3.4 Extensions to a Larger Number of Factors

The ANOVA model can, in principle, be extended to take into account any number of qualitative factors. In this situation, the full ANOVA model also includes all possible interactions of higher order. If, for example, one considers a model with four factors, then the full ANOVA model includes the main effects for all four factors, six two-way interactions between the pairs of factors, four three-way interactions describing the joint influence of any subset of three factors, and one four-way interaction between all factors. It is often the case that such a full ANOVA model is so complex that the sample size might not be large enough to estimate all the possible interactions. Therefore, in practice, ANOVA models with a large number of factors are often reduced by only considering main effects, or perhaps just including interactions of small order.

8.3.5 Multiple Regression

Multiple regression is used to investigate the functional relationship between a continuous variable Y and m numerical variables $X_1, \ldots, X_m$, where a linear relationship is assumed. Given $n > m + 1$ independent realizations of the vector $(X_1, \ldots, X_m)$ (although X_i may be associated with X_j), this relationship takes the form

$$Y_i = \beta_0 + \sum_{j=1}^{m} \beta_j x_{ij} + \varepsilon_i, \tag{8.9}$$

or in vector form

$$Y = X\beta + \varepsilon. \tag{8.10}$$

The dependent variable $Y = (Y_1, \ldots, Y_n)^T$ in this model is often called the regressand, while the m explicatory variables $X_j = (x_{1j}, \ldots, x_{nj})^T, j = 1, \ldots, m$, are called regressors. In vector form, the matrix X is defined as $X = (\mathbb{1}, X_1, \ldots, X_m)$, where $\mathbb{1}$ is a vector containing just n ones and for $i > 0$ the component β_i of the vector $\beta = (\beta_0, \beta_1, \ldots, \beta_m)^T$ is the effect size of the ith regressor, i.e., when the value of the ith regressor increases by one, then, according to the model, the value of the regressand Y increases on average by β_i. The component β_0 is the constant in the regression equation. The usual minimal assumptions on the random error term $\varepsilon = (\varepsilon_1, \ldots, \varepsilon_n)^T$ are that ε_i are i.i.d. and $E(\varepsilon_i) = 0$.

Classical regression theory uses the least squares estimate $\hat{\beta}$ of the coefficient vector β. This estimate minimizes the residual sum of squares

$$RSS = (Y - X\hat{\beta})^T (Y - X\hat{\beta}) = \sum_{i=1}^{n} (Y_i - \hat{Y}_i)^2, \tag{8.11}$$

where $\hat{Y} = (\hat{Y}_1, \ldots, \hat{Y}_n)^T = X\hat{\beta}$. It can be shown that whenever the matrix X has full rank, $m + 1$, then $\hat{\beta} = (X^T X)^{-1} X^T Y$.

One major task in regression analysis is to test linear hypotheses of the form $C\beta = \xi$, where C is a $s \times (m + 1)$ matrix with full rank s, and the components of ξ, a vector of length s, are commonly all equal to 0. The key ingredient in defining such tests is again a decomposition of the sum of squares, where one estimates $\hat{\beta}_C$ under the constraint that $C\beta = \xi$, and then defines

$$RSS_C = (Y - X\hat{\beta}_C)^T (Y - X\hat{\beta}_C).$$

Testing linear hypotheses regarding multiple regression models:

Given model (8.9) and the $\varepsilon_i \sim \mathcal{N}(0, \sigma^2)$ are i.i.d., the usual test statistic for the linear hypothesis $H_0 : C\beta = \xi$ is given by

$$F = \frac{(RSS_C - RSS)/s}{RSS/(n - m - 1)},$$

and under H_0 this test statistic is F-distributed with s and $n - m - 1$ degrees of freedom, $F \sim F_{s,n-m-1}$.

The most common linear hypothesis is concerned with the question of whether there is any linear relationship between Y and the regressors X_j at all, which is formulated as $H_0 : \beta_1 = \cdots = \beta_m = 0$, or in vector form

$$C\beta = 0, \quad \text{where} \quad C = \begin{pmatrix} 0 & 1 & & 0 \\ \vdots & & \ddots & \\ 0 & 0 & & 1 \end{pmatrix}. \tag{8.12}$$

Obviously, in this case $s = m$. The second most common type of hypothesis is to test whether an individual regressor is of importance, i.e., $H_0 : \beta_j = 0$. Given the model (8.9), the matrix describing this linear hypothesis is simply $C = (0 \ldots 0 \, 1 \, 0 \ldots 0)$, a row vector of zeros except for the $j + 1$th component. The corresponding F-Test with $s = 1$ can be shown to be equivalent to the frequently used t-test for this hypothesis.

For details and further material on multiple linear regression, we again refer the reader to [8]. The traditional representation given here relies upon least squares estimates and a decomposition of the sum of squares. For linear regression with normally distributed error terms, the F-tests described above are closely related to likelihood ratio tests. The maximum likelihood approach is discussed in Sect. 3.3.1.

8.3.6 Weighted Least Squares

Now assume that data are generated according to the model

$$Y_i = \beta_0 + \sum_{j=1}^{m} \beta_j x_{ij} + \varepsilon_i, \tag{8.13}$$

where $\varepsilon_i \sim \mathcal{N}(0, \sigma_i^2)$.

In this case, estimating the vector of regression coefficients $(\beta_1, \ldots, \beta_m)$ using the method of maximum likelihood reduces to the problem of minimizing the following weighted sum of squared residuals:

$$\hat{\beta} = \operatorname{argmax}_\beta \sum_{i=1}^{n} (Y_i - \sum_{j=1}^{m} \beta_j x_{ij})^2 / \sigma_i^2, \tag{8.14}$$

where the ith squared residual is weighted by $w_i = \frac{1}{\sigma_i^2}$. Thus, data points associated with an error term of small variance have a stronger influence on the estimated regression line than points associated with an error term of large variance, which typically lie further away from the regression line. It can be shown that the solution to this optimization problem (8.13) is

$$\hat{\beta} = (X^T W X)^{-1} X^T W Y, \tag{8.15}$$

where W is the diagonal matrix of dimension $n \times n$ with elements $w_{ii} = \frac{1}{\sigma_i^2}$.

8.4 General Linear Models

We discussed ANOVA and multiple linear regression separately. The powerful approach of general linear models (GLMs) combines both types of analysis into a single framework. This is achieved by defining ANOVA within a regression setting. To this end, an appropriate coding of the various group variables is necessary, where the conceptually simplest coding makes use of dummy variables. Consider one-way ANOVA where the factor has k groups and in the rth group there are n_r observations $y_{r1}, \ldots, y_{rn_r}$. Stack all the observations into one vector $Y = (y_{11}, \ldots y_{1n_1}, \ldots, y_{k1}, \ldots y_{kn_k})^T$. Then for $r \in \{1, \ldots, k-1\}$ define the n_r components of the vector X_r of length n corresponding to observations in group r to be equal to 1, the n_k components corresponding to observations in group k to be equal to -1 and all the other components to be 0. The $n \times k$ matrix $X = (\mathbb{1}, X_1, \ldots, X_{k-1})$ is called the design matrix. Then the regression model $Y = X\beta$ with $\beta = (\beta_0, \beta_1, \ldots, \beta_{k-1})^T$ is equivalent to one-way ANOVA. This regression model states that for $r \in \{1, \ldots, k-1\}$ the mean in group r is $\beta_0 + \beta_r$ and the mean in group k is $\beta_0 - \sum_{j=1}^{k-1} \beta_j$. Hence, the null hypothesis $\beta_1 = \beta_2 = \cdots = \beta_{k-1} = 0$ is equivalent to the null hypothesis that all the group means are equal to β_0. Thus the ANOVA F-test described in Table 8.1 corresponds to the test of the null hypothesis that not all the regression coefficients are equal to zero (8.12). Since a qualitative factor with k levels needs to be described by $k - 1$ columns in the design matrix, the number of degrees of freedom for a one-way ANOVA model is equal to $k - 1$. The interactions between different qualitative factors can be represented by dummy variables obtained by taking the products of the dummy variables coding for the main effects of these factors. Thus the number of dummy variables coding for a two-way interaction between two factors with I and J levels, respectively, is equal to $(I - 1)(J - 1)$.

Creating design matrices for categorical variables enables us to apply the powerful machinery of linear regression not only for one-way ANOVA, but for any number of factors. Moreover, predictors that are categorical variables can easily be combined with continuous regressors, as in covariance analysis (ANCOVA).

The method of creating dummy variables for qualitative predictors, described above, is just one of many possibilities. Actually, one can use any method of coding, such that for each i the dummy variable X_i takes the same value for all individuals from the same group and the design matrix $X = (\mathbb{1}, X_1, \ldots, X_{k-1})$ is of full order. While the interpretation of the regression coefficients obviously depends on the chosen method, the values of the F-test statistics for the main and interaction effects remain unchanged, irrespectively of the method of coding.

8.4.1 Cockerham's Model

Here, we will briefly discuss Cockerham's model, which is a specific method of defining dummy variables within the framework of QTL mapping to apply a regression model. We consider the intercross $F2$ design discussed in Sect. 2.2.1.2. For a given genetic locus, the two dummy variables, X and Z, used in Cockerham's model are specified in Table 2.3.

For a regression model of the form

$$E(Y) = \mu_0 + \beta X + \gamma Z,$$

it holds that

$$\beta = [E(Y|aa) - E(Y|AA)]/2, \text{ and } \gamma = E(Y|Aa) - \frac{E(Y|AA) + E(Y|aa)}{2}.$$

From the arguments presented in Sect. 2.2.2, it follows that, according to Cockerham's model, β can be interpreted as the additive effect and γ as the dominance effect.

Another important property of Cockerham's model is that in an ideal $F2$ population the random variables X and Z are not correlated. Therefore, their effects can be estimated separately from each other. Also, Cockerham's coding can be used to build a multiple regression model which considers the influence of many loci and their interactions. As usual, the dummy variables corresponding to the interactions are the products of the dummy variables corresponding to the main effects of the genes involved in a given interaction. Thus, using an $F2$ design, the interaction between a pair of genes can be broken into four components, namely X_1X_2, X_1Z_2, Z_1X_2, and Z_1Z_2. If the loci under consideration are not linked to each other, then the dummy variables corresponding to different components of such a regression model are not correlated either, which avoids the confounding of errors when estimating various genetic effects.

8.4.2 Robustness of General Linear Models

Multiple linear regression and ANOVA models assume that the original samples are drawn from normal populations. However, from the central limit theorem, the significance levels of F-tests are close to their nominal levels as long as the distribution generating the data Y has a finite variance and the sample size is large enough. Unless the distribution generating the data has "heavy tails," a sample size exceeding 100 usually suffices for this approximation to be accurate. The presence of outliers is an indication that the application of standard linear models is inappropriate.

8.5 Generalized Linear Models

Generalized linear models [5] are used to describe the relationship between the response variable and the explanatory variables in the case when the distribution of the response has a specific form, different from the normal distribution. Typical examples include logistic regression to model binary traits, or Poisson regression when the dependent trait is a count variable. Typically, generalized linear models are abbreviated to GLMs, just as general linear models are. To make the distinction clear, in this book we will use the abbreviation gLMs for generalized linear models.

Let $Y = (Y_1, \ldots, Y_n)^T$ be a vector of n trait values. Using a generalized linear model (gLM), the density of Y_i (for continuous traits) or the probability mass function of Y_i (for discrete traits) stems from an overdispersed exponential family, which means that it can be written in a very specific form.

The overdispersed exponential family:

When a random variable comes from the overdispersed exponential family, the density (or probability mass function, as appropriate) can be written as

$$f(y_i; \theta_i, \phi) = \exp\left\{\frac{\theta_i y_i - \psi(\theta_i)}{\phi}\right\} h(y_i, \phi), \qquad (8.16)$$

where θ_i is called the canonical parameter and ϕ the dispersion parameter.

Note that this family is a generalization of the (regular) exponential family. A distribution belongs to the exponential family when the dispersion parameter is given by a constant.

The canonical parameter θ_i is usually unknown. The dispersion parameter ϕ may be known or unknown, while the functions h and ψ have to be known. For such a distribution, it can be shown that $\mu(\theta_i) := E(Y_i) = \psi'(\theta_i)$ and $\mathrm{Var}(Y_i) = \phi \psi''(\theta_i)$.

First note that clearly the normal distribution (6.1) is a member of the overdispersed exponential family, because

$$f(y_i; \mu_i, \sigma) = \frac{1}{\sqrt{2\pi}\sigma} \exp\left(-\frac{(y_i - \mu_i)^2}{2\sigma^2}\right) = \exp\left(\frac{y_i \mu_i - \mu_i^2/2}{\sigma^2}\right) \frac{\exp\left(-y_i^2/2\sigma^2\right)}{\sqrt{2\pi}\sigma}.$$

Here, the canonical parameter equals the mean, $\theta_i = \mu_i$, and the dispersion parameter is simply the variance, $\phi = \sigma^2$. The other two functions are readily identified as $\Psi(\theta_i) = \theta_i^2/2$, and $h(y_i, \phi) = \frac{1}{\sqrt{2\pi\phi}} \exp\left(-y_i^2/2\phi\right)$.

Perhaps the most important gLM is **logistic regression**, where the dependent trait has a binomial distribution, $Y_i \sim B(1, \pi_i)$. The corresponding probability mass function (6.10) can be rewritten as

$$f(y_i; \pi_i) = \pi_i^{y_i}(1 - \pi_i)^{1-y_i} = \exp\left\{y_i \log\frac{\pi_i}{1 - \pi_i} + \log(1 - \pi_i)\right\}.$$

Here we have $\phi = 1$, thus $B(1, \pi_i)$ is a member of the regular exponential family (without overdispersion). The canonical parameter is given by $\theta_i = \log\frac{\pi_i}{1-\pi_i}$ and, furthermore, we have $h(y_i, \phi) = 1$, and $\psi(\theta_i) = \log(1 + \exp\{\theta_i\})$.

In the case of **Poisson regression**, sometimes referred to as the log-linear model, count data is modeled by a Poisson distribution, $Y_i \sim P(\lambda_i)$. The corresponding probability mass function (6.11) can be rewritten as

$$f(y_i; \lambda_i) = \frac{\exp\{-\lambda_i\}}{y_i!} \lambda_i^{y_i} = \frac{1}{y_i!} \exp\{y_i \log\lambda_i - \lambda_i\}.$$

Again, we have $\phi = 1$, which shows that the Poisson distribution is also a member of the regular exponential family. The canonical parameter is $\theta_i = \log \lambda_i$ and, furthermore, we have $h(y_i, \phi) = \frac{1}{y_i!}$, and $\psi(\theta_i) = \exp\{\theta_i\}$.

Let $X_i = (1, X_{i1}, \ldots, X_{im})^T$ denote the column vector which contains the values of the m observations of the explanatory variables for the i-th individual, and let $X = (X_1, \ldots, X_n)^T$ be the corresponding design matrix. The fundamental principle underlying gLMs is that the expected values of the traits are given as a function of the canonical parameters, $\mu(\theta) = (\mu(\theta_1), \ldots, \mu(\theta_n))$, for which the following relationship holds:

The generalized linear model:

$$g(\mu(\theta_i)) = X_i^T \beta, \qquad (8.17)$$

where $\beta = (\beta_0, \beta_1, \ldots, \beta_m)^T$ is a vector of (unknown) regression coefficients and g is called the **link function**.

This looks almost like the relationship described by linear regression (8.10), except for the link function g. In principle, the link function could be any monotonic differentiable function (for details see [5]). However, depending on the distribution of the trait in question, there are certain choices which are commonly used. In particular, the link function $g = \mu^{-1}$ is called canonical and yields the relationship $\theta_i = X_i\beta$.

Considering the three examples given above, the canonical link function for the normal distribution is simply the identity map and the appropriate gLM corresponds to linear regression. Thus GLMs are a special case of gLMs. In the case of logistic regression, one has $\mu(\theta_i) = \psi'(\theta_i) = \frac{\exp\{\theta_i\}}{1+\exp\{\theta_i\}}$ and the canonical link function is given by $g(z) = \mu^{-1}(z) = \log \frac{z}{1-z}$. This is the well-known logit function, which for $z = \pi_i$ corresponds to the logarithm of the odds ratio for the probability π_i (compare with Eq. (3.22) in Sect. 3.3.1). In the case of log-linear (Poisson) regression, $\mu(\theta_i) = \exp\{\theta_i\}$ and the canonical link function is given by $g(z) = \log(z)$.

Now, consider the problem of testing the hypothesis that some of the explanatory variables, say $X_1, \ldots, X_k$ where $k \leq m$, have no influence on the trait Y. This problem can be formalized as the problem of testing $H_0 : (\beta_1, \ldots, \beta_k) = \mathbf{0}$ against the alternative $H_A : (\beta_1, \ldots, \beta_k) \neq \mathbf{0}$. Let M_0 denote the reduced model specified by the null hypothesis, i.e., the model where the coefficients $\beta_1, \ldots, \beta_k$ are all set to 0. One of the common tests of H_0 is the likelihood ratio test with test statistic given by

$$LRT_j = -2\log \frac{L(Y; \hat{\beta}_R, \hat{\phi}_R)}{L(Y; \hat{\beta}, \hat{\phi})}, \qquad (8.18)$$

where

$$L(Y; \beta, \phi) = \prod_{i=1}^{n} f(y_i; \theta_i(\beta, X_i), \phi) \tag{8.19}$$

denotes the likelihood of the data, $\hat{\beta}$ and $\hat{\phi}$ denote the maximum likelihood estimates of β and ϕ according to the full gLM model and $\hat{\beta}_R$, $\hat{\phi}_R$ are the corresponding maximum likelihood estimates according to the reduced model M_0 (see also Sect. 3.3.1).

Under some regularity conditions on the density function f and the design matrix $X_{n \times m}$, the test statistic LRT_j has asymptotically (as $n \to \infty$) a chi-square distribution with k degrees of freedom (see for example Theorem 6.7 of [9]). The regularity conditions on the density function are satisfied by most popular gLMs, including logistic regression and Poisson regression. The regularity conditions on the design matrix are needed to guarantee that for large n this matrix is non-singular.

8.5.1 Extensions of Poisson Regression

To model the dependence of a count response variable on explanatory variables, one can also use a zero-inflated generalized Poisson regression model (ZIGPR), introduced in [1]. In this model, the distribution of the trait given values of predictors is modeled by the zero-inflated Poisson distribution, $Y_i \sim ZIGP(\mu_i, \varphi, \omega_i)$ (see 6.13), where μ_i and ω_i depend on regressors through the log-linear and logit link functions, respectively

$$\log \mu_i = \beta_0 + \sum_{j=1}^{k} \beta_j X_{ji},$$

$$\frac{\exp\{\omega_i\}}{1 + \exp\{\omega_i\}} = \gamma_0 + \sum_{j=1}^{k} \gamma_j X_{ji}.$$

The class of ZIGPR models contains the subclasses of zero-inflated Poisson regression (ZIPR, $\varphi = 1$), generalized Poisson regression (GPR, $\omega = 0$) and standard Poisson regression (PR, $\varphi = 1$, $\omega = 0$).

8.6 Linear Mixed Models

We will briefly sketch the theory of linear mixed models (LMMs), which in general play an important role in statistical genetics and are used, in particular, in the context of modeling population structure in GWAS. For a more comprehensive treatment of LMMs we recommend [4].

We have already seen in Sect. 8.4 how to perform ANOVA in the context of GLMs, where a model is simply stated as $Y = X\beta + \varepsilon$ and X is an appropriate design matrix coding the factors to be analyzed. The formal definition of a linear mixed model looks quite similar

Linear Mixed Model (LMM):

A linear mixed model for a dependent variable Y based on N sets of observations can be expressed in the form

$$Y = X\beta + Zu + \varepsilon, \tag{8.20}$$

where $\beta \in \mathbb{R}^k$ are the fixed effects, $u \sim \mathcal{N}_q(0, D)$ the random effects, and $X \in R^{N \times k}$ and $Z \in R^{N \times q}$ are the corresponding design matrices. The error term is also typically assumed to be normally distributed, $\varepsilon \sim \mathcal{N}_N(0, R)$ and independent of u.

An important special case of an LMM is simple *one-way random effects ANOVA*, which includes no fixed effects apart from the intercept and one random effect corresponding to a categorical variable with q factors. Consequently, Z is simply the design matrix of a one-way ANOVA procedure which is typically coded using dummy variables, $Z = (Z_1, \ldots, Z_q)$, where $Z_{ij} = 1$ if the jth observation belongs to group i, otherwise $Z_{ij} = 0$. The simplest choice of the variance terms is then $D = \sigma_Z^2 I_q$ and $R = \sigma_\varepsilon^2 I_N$ (where I_n is the $n \times n$ identity matrix) and there remain only three parameters to be estimated, β_0, σ_Z^2 and σ_ε^2. Using LMMs, parameter estimation is usually based on maximum likelihood methods, where the notion of restricted maximum likelihood (REML) is discussed below.

There are two different ways of interpreting LMMs. First, the form of Eq. (8.20) emphasizes the difference between fixed effects and random effects. Fixed effects representing categorical variables play exactly the same role as factors in ANOVA, for which individual group means can be estimated. Random effects also typically comprise categorical variables, which are coded by the design matrix Z. However, in this case, the group means are not estimated individually, but as realizations of a random variable. For example, in the case of a one-way random effects model, one does not have to estimate q parameters describing the group means, but only one parameter σ_Z^2, which models the variance of the random effect. Individual group means are not estimated directly, but they can be predicted from the resulting LMM model.

The second way of looking at LMMs is related to the fact that Eq. (8.20) actually serves the purpose of defining a linear model with a very specific covariance structure. In fact, it is easy to see that under (8.20), $V := \text{Cov}(Y) = ZDZ' + R$ and thus

$$Y \sim \mathcal{N}(X\beta, \, ZDZ' + R).$$

This point of view is emphasized when it comes to using an LMM to model the population structure in GWAS. If the covariance matrix V is known, then the ML estimates of the fixed effects are given by

$$\hat{\beta} = (X'V^{-1}X)^{-1}X'V^{-1}Y. \tag{8.21}$$

For the sake of simplicity, we assume that the design matrix X has full rank; otherwise we would have to work with generalized inverse matrices (see [4] for more details). In the case of the natural situation where V is unknown, the ML estimates $\hat{\beta}$ are computed in exactly the same way, only V is substituted by its estimate $\hat{V}$.

There are various methods available to estimate V. As in ANOVA, there exists an approach based on decomposing the sum of squares, but this approach only really works for balanced designs (where the number of observations corresponding to each combination of factors is constant) and seems to be mainly of historical interest today, as ML methods are employed almost exclusively. Except for in the very simplest situations, such as one-way random effects ANOVA with a balanced design, no closed formulas for ML estimates of V are available. Therefore, one has to rely upon numerical optimization methods, such as the Newton–Raphson algorithm or numerically more stable refinements thereof.

One of the most widely used methods of estimating V is based on restricted maximum likelihood (REML), where one considers all the linear contrasts of Y which are orthogonal to the design matrix of the fixed effects X. More formally, if X has rank k, one utilizes an $N \times (N - k)$ matrix $C = [c_1, \ldots, c_{N-k}]$ with linearly independent columns and $C'X = 0$. The REML estimate of V is then obtained by maximizing the likelihood of $C'Y$. It turns out that the exact choice of C does not have an influence on the estimate obtained. REML has certain theoretical properties which make it preferable to ML when estimating V. However, for large sample sizes both methods tend to give very similar results. Thus, in the case of GWAS, the choice of ML or REML estimation of the covariance matrix does not seem to be a big issue.

Having estimated both the fixed effects β and the covariance structure of the LMM (i.e., D and R), one is often interested in learning about the random effect u. To this end, one wishes to find a predictor of u which has some kind of optimality properties. In general, if one has two random vectors $u \in \mathbb{R}^q$ and $Y \in \mathbb{R}^N$ with joint density $f(u, Y)$ and assumes that only y is known, then one looks for a predictor of u as a function of Y, $\tilde{u} = \tilde{u}(Y)$. One criterion used to find the best predictor is to assess $\tilde{u}$ according to its mean squared error, where

$$MSE := E(\tilde{u} - u)^2 = \int_y \int_u (\tilde{u} - u)^2 f(u, y) dy du,$$

which is minimized by setting $\tilde{u}$ equal to the conditional expected value $E(u|Y)$. It can be shown that given a specific model of the form

$$\begin{pmatrix} u \\ Y \end{pmatrix} \sim \mathcal{N} \left[\begin{pmatrix} 0 \\ X\beta \end{pmatrix}, \begin{pmatrix} D & DZ' \\ ZD & V \end{pmatrix} \right],$$

then the best predictor (BP) coincides with the best linear predictor (BLP) and is of
the form

$$\tilde{u} = DZ'V^{-1}(Y - X\beta).$$

Obviously, in practice, β is not known in advance, but is estimated by $\hat{\beta}$ according to
Eq. (8.21) with V being substituted by $\hat{V}$. If we additionally assume that $\tilde{u}$ should be
unbiased, in the sense that $E\tilde{u} = 0$, then the **best linear unbiased predictor** (BLUP)
has the form

$$\tilde{u} = DZ'V^{-1}(Y - X\hat{\beta}) . \tag{8.22}$$

The **Henderson equations** give an extremely elegant way of simultaneously computing the BLUE (best linear unbiased estimator) and BLUP for given (or estimated)
R and D matrices

$$\begin{pmatrix} X'R^{-1}X & X'R^{-1}Z \\ Z'R^{-1}X & Z'R^{-1}Z + D^{-1} \end{pmatrix} \begin{pmatrix} \tilde{\beta} \\ \tilde{u} \end{pmatrix} = \begin{pmatrix} X'R^{-1}y \\ Z'R^{-1}y \end{pmatrix}.$$

8.7 Nonparametric Tests

All the tests discussed so far have been parametric tests. This means that they are based
on a parametric model with clearly specified assumptions regarding the underlying
probability distribution of the trait of interest. For example, an important assumption
in ANOVA and regression models is that the conditional distribution of the dependent
trait given the values of the explanatory variables is normal, leading to the test statistic
having an F-distribution under the null hypothesis. From the central limit theorem,
for large sample sizes some deviation from these basic assumptions is often tolerable.
However, if the sample size is not sufficiently large, and especially if the distribution
of the dependent trait is clearly non-normal, then the use of classical parametric tests
can become quite problematic. Depending on the kind of non-normality involved,
either the type I error or the type II error of the testing procedure can become inflated.

   One way to proceed when the assumptions of a parametric model are violated is
to apply a nonparametric test. Permutation tests and resampling procedures belong
to this category of tests and are discussed in Sect. 3.2.2 in the context of multiple
testing. This section briefly recalls some fairly simple nonparametric alternatives to
the two sample t-test, ANOVA and multiple regression. For an extensive treatment
of rank based nonparametric methods, we refer the reader to [2].

   The two sample t-test is performed to check whether the mean of a variable is
identical for two different groups. When the variances in both groups are identical,
then the test statistic (8.4) is used, where it is assumed that observations are independent of each other and normally distributed. If the normality assumption is violated,
one can use the Wilcoxon rank-sum test, or equivalently the Mann–Whitney U test,
instead of the two sample t-test. Here, the distribution of the data can be completely
general, but it is assumed that the distribution is the same in both groups up to a shift

in location. In terms of the cumulative distribution functions (cdf), this means that $F_2(x) = F_1(x - a)$, where F_i is the cdf for group i. The Wilcoxon rank-sum test can be used to test the null hypothesis $H_0 : a = 0$. Note that this hypothesis can be interpreted as either stating that the means are same for each group or that the medians are the same. The test statistic is obtained in the following way:

Wilcoxon rank-sum test:

Let X_i be i.i.d. for $i = 1, \ldots, n_1$ with cdf F_1 and Y_j i.i.d. for $j = 1, \ldots, n_2$ with cdf F_2, where $F_2(x) = F_1(x - a)$. Define the ranks of X_i in the pooled data to be $R_i = \text{Rank}(X_i)$, $i = 1, \ldots, n_1$. To test $H_0 : a = 0$, we use the test statistic

$$W = \sum_{i=1}^{n_1} R_i. \tag{8.23}$$

For small sample sizes, n_1 and n_2, the p-values for each possible value of W have been computed using combinatorial methods and can be found in tables. For larger sample sizes, normal approximations to the distribution of W are appropriate. For a detailed exposition, see [2]. If the data are actually normally distributed, then the Wilcoxon rank-sum test is slightly less powerful than the two sample t-test. On the other hand, the Wilcoxon test is more robust to outliers and can also be applied when the variable of interest is ordinal (i.e., categorized according to some scale).

Rank-based methods can also be used to compare medians for more than two groups. Consider again the one-way ANOVA setting with k groups, where n_i observations $y_{ij}, j = 1, \ldots, n_i$ are observed for group i. As in Sect. 8.3.1, the total number of observations is given by $N = \sum_{i=1}^{k} n_i$. Let the rank of y_{ij} among all N observations be denoted as R_{ij}.

The Kruskal–Wallis test:

Given the ranks R_{ij} of observations $j \in \{1, \ldots, n_i\}$ of group $i \in \{1, \ldots, k\}$, define $\bar{R}_{i.} = \frac{1}{n_i} \sum_{j=1}^{n_i} R_{ij}$. The Kruskal–Wallis test statistic is defined as follows:

$$K = \frac{12}{N(N+1)} \sum_{i=1}^{k} n_i \left(\bar{R}_{i.} - (N+1)/2 \right)^2. \tag{8.24}$$

The test statistic (8.24) can be written as $K = (N - 1)MSS_R/TSS_R$, where the sum of squares explained by the model, MSS_R and the total sum of squares TSS_R are based on the ranks R_{ij}. This relationship follows from the facts that $\bar{R} = (N + 1)/2$, and $TSS_R = (N - 1)N(N + 1)/12$. For very small sample sizes, it is again possible to compute exact p-values using combinatorial arguments. For larger sample sizes, the test statistic K is approximately χ^2-distributed with $k - 1$ degrees of freedom (see [2]).

8.7.1   Wilcoxon Signed-Rank Test

A rank test can also be used for a two sample paired design. As in Sect. 8.2.3, let us denote the measurements in the first sample by $X_1, \ldots, X_n$ and the measurements in the second sample by $Y_1, \ldots, Y_n$. We assume that for each i the two measurements are paired, as in the example of two sequential measurements made on one group of patients. As in the Wilcoxon rank sum test, our aim is to decide whether the probability distributions generating these two samples differ by a location parameter. To this end, we calculate the differences $Z_i = Y_i - X_i$ and then compute the vector of the ranks $(R_1, \ldots, R_n)$ for the absolute values of the differences $|Z_1|, \ldots, |Z_n|$. Finally, we calculate the following test statistic:

Wilcoxon signed-rank test:

$$W_S = |\sum_{i=1}^{n} \text{sgn}(Y_i - X_i)R_i|.$$

The null hypothesis of no difference between the two distributions is rejected for large values of W_S. As in the case of the Wilcoxon rank-sum test, when the sample size n is small, critical values for W_S can be found in statistical tables. For larger sample sizes, normal approximations for the distribution of W_S are available.

8.7.2   Rank Regression

Rank methods can also be used to develop a nonparametric version of multiple regression. Let $R = (R_1, \ldots, R_n)$ denote the vector of ranks corresponding to n observed trait values $Y_1, \ldots, Y_n$. We want to determine how (and whether) the trait depends on k predictors. Let $X_{n \times k}$ denote the corresponding design matrix (without an intercept) and define $X_{i.} = (X_{i1}, \ldots, X_{ik})$. The nonparametric multiple regression model looks very similar to (8.9), but is based on the minimal assumption that for $i = 1, \ldots, n$ the error terms $\varepsilon_i = Y_i - \sum_{j=1}^{k} \beta_j X_{ij}$ are i.i.d. random variables from the same continuous distribution. The exact form of this distribution can remain unspecified.

The null hypothesis that the predictors have no influence on the trait Y can be written in the familiar form $H_0 : \beta_1 = \cdots = \beta_k = 0$. This can be tested using the following rank statistic

Test statistic for rank regression:

$$T = \frac{12}{n(n+1)} L^T C^{-1} L,$$

where $L = (L_1, \ldots, L_k)$, $L_j = \sum_{i=1}^{n} R_i(X_{ij} - \bar{X}_{.j})$, $C = \sum_{i=1}^{n}(X_{i.} - \bar{X})^T$ $(X_{i.} - \bar{X})$ and $\bar{X} = \frac{1}{n} \sum_{i=1}^{n} X_{i.}$.

According to Theorem 5.3.1 of [7], under some standard restrictions on the design matrix, for $n \to \infty$ the distribution of the test statistic T converges to a chi-square distribution with k degrees of freedom.

8.8 Tests for Categorical Variables

With the exception of gLMs, we have so far mainly discussed examples of statistical testing where the response variable is quantitative. Here, we will focus on chi-square tests that can be used when the response variable is categorical. We will also briefly discuss Fisher's exact test, which is very useful for small sample sizes.

8.8.1 Chi-Square Goodness-of-Fit Test

The chi-square goodness-of-fit test is concerned with the question of whether the observed frequencies of some qualitative variable are in accordance with the probabilities specified by a given discrete distribution. For example, consider the question of segregation distortion at some locus in an $F2$ population. This problem can be described as follows: If no segregation distortion occurs, then one expects the relative frequencies of the genotypes aa, aA, AA in the population to be close to the theoretical probabilities, 1/4, 1/2, and 1/4, respectively. Let p_1, p_2, p_3 denote the true population probabilities for these three genotypes. Then the hypothesis of no segregation distortion can be written as $H_0 : p_1 = 1/4, p_2 = 1/2, p_3 = 1/4$. Based on the observed frequencies, one would like to decide whether the null hypothesis is violated, in which case one would conclude that segregation is distorted.

In general, consider a categorical variable X which can take k different values. It is customary to code these values using the natural numbers $\{1, \ldots, k\}$ and to write $p_i = P(X = i)$. We are interested in testing the hypothesis that the vector of probabilities $p = (p_1, \ldots, p_k)$ is equal to some fixed prespecified vector $p_0 = (p_{10}, \ldots, p_{k0})$

The goodness-of-fit hypothesis:

$$H_0 : (p_1, \ldots, p_k) = (p_{10}, \ldots, p_{k0}). \tag{8.25}$$

We will test the null hypothesis (8.25) based on a sequence $X_1, \ldots, X_n$ of independent observations of our categorical variable. Let $e_i = np_{i0}$ denote the expected number of occurrences of the ith value of our variable under the hypothesis H_0 and n_i denote the actual number of times this value was observed. The hypothesis H_0 is rejected for large realizations of the following test statistic:

Chi-square goodness-of-fit test:

The null hypothesis in a goodness-of-fit test can be tested using the statistic

$$T = \sum_{i=1}^{k} \frac{(n_i - e_i)^2}{e_i}.$$

Assuming that $p_{i0} \in (0, 1)$ for each i, then under the null hypothesis the distribution of T converges for large n towards a chi-square distribution with $k - 1$ degrees of freedom. To apply this approximation, it is usually required that the expected number of observations $e_i \geq 5$ for each i. Many statistical programs carry out "exact" chi-square goodness-of-fit tests, which rely on the application of Monte Carlo methods to estimate p-values. A slightly different variant of a goodness-of-fit test is used in Sect. 5.3 to test whether genotype frequencies correspond to the Hardy–Weinberg equilibrium.

8.8.2   Chi-Square Test of Independence

Another prominent example of a chi-square test is the test of independence for two categorical variables. Consider two genetic markers with alleles A/a and B/b, respectively. One might be interested in whether the genotypes of these allelic markers are independent of each other. As above, the chi-square test statistic compares the

Table 8.5 Example of a χ^2-test of independence

	Observed					Expected			
	AA	Aa	aa	RowSum		AA	Aa	aa	RowSum
BB	n_{11}	n_{12}	n_{13}	$n_{1.}$	BB	e_{11}	e_{12}	e_{13}	$n_{1.}$
Bb	n_{21}	n_{22}	n_{23}	$n_{2.}$	Bb	e_{21}	e_{22}	e_{23}	$n_{2.}$
bb	n_{31}	n_{32}	n_{33}	$n_{3.}$	bb	e_{31}	e_{32}	e_{33}	$n_{3.}$
ColSum	$n_{.1}$	$n_{.21}$	$n_{.3}$	N	ColSum	$n_{.1}$	$n_{.21}$	$n_{.3}$	N

observed frequencies of each genotype combination with their expected frequencies under the assumption of independence. This is illustrated in Table 8.5.

$$e_{ij} = \frac{n_{i.}n_{.j}}{n}, \qquad n_{i.} = \sum_{j=1}^{3} n_{ij}, \quad n_{.j} = \sum_{i=1}^{3} n_{ij} \quad N = \sum_{i=1}^{3}\sum_{j=1}^{3} n_{ij}. \qquad (8.26)$$

χ^2-test of independence:

Consider two categorical variables X and Y with r and s categories, respectively. Let n_{ij} be the number of observations where $X = i$ and $Y = j$, and let the expected counts under independence, e_{ij}, be defined as in (8.26). Then, under the assumption that X and Y are independent, the test statistic

$$T = \sum_{i=1}^{r}\sum_{j=1}^{s} \frac{(n_{ij} - e_{ij})^2}{e_{ij}}$$

asymptotically has a χ^2-distribution with $(r-1)(s-1)$ degrees of freedom, $T \sim \chi^2_{(r-1)(s-1)}$.

As in the case of testing goodness-of-fit, the appropriate approximation should be applied only when the expected number of observations in each cell exceeds some prespecified value (often $e_{ij} \geq 5$ is recommended, although this choice is rather conservative). Otherwise, one can use statistical software to carry out "exact" chi-square tests of independence, which use Monte Carlo methods to estimate p-values.

8.8.3 Fisher's Exact Test

Fisher's exact test is often used as an alternative to the chi-square test of independence when the sample size is too small to apply the chi-square approximation. Here, we will explain the principle of this test only in the case of 2×2 contingency tables, which means that both categorical variables can take exactly two values. However, in many statistical software packages one can also carry out this test for larger contingency tables.

Assume that for a 2×2 table the marginal frequencies $n_{1.}$, $n_{2.}$ and $n_{.1}$ are known. Then n and $n_{.2}$ can be derived in a simple manner and knowing one cell count (for example n_{11}) determines the whole table. Fisher showed that, under the assumption of independence, the probability of observing a cell count n_{11} given $n_{1.}$, $n_{2.}$ and $n_{.1}$ follows the following hypergeometric distribution:

$$P(n_{11}) = \frac{\binom{n_{1.}}{n_{11}}\binom{n_{2.}}{n_{21}}}{\binom{n}{n_{.1}}}.$$

The p-value for Fisher's exact test is subsequently calculated by summing up the probabilities of all cell counts n_{11} which are at least as extreme (compared to the expected value under the null hypothesis) as the observed value. One can consider one-sided or two-sided hypotheses. The null hypothesis is rejected if the p-value does not exceed the assumed significance level.

References

1. Famoye, F., Singh, K.P.: Zero-inflated generalized Poisson model with an application to domestic violence data. J. Data Sci. **4**, 117–130 (2006)
2. Lehmann, E.L., D'Abrera, H.J.M.: Nonparametrics: Statistical Methods Based on Ranks. McGraw-Hill, New York (1975)
3. Lehmann, E.L., Romano, J.P.: Testing Statistical Hypotheses. Springer, New York (2005)
4. McCulloch, C.E., Searle, S.R., Neuhaus, J.M.: Generalized, Linear and Mixed Models, 2nd edn. John Wiley and Sons (2008)
5. Nelder, J.A., McCullagh, P.: Generalized Linear Models. Chapman and Hall, London (1991)
6. Neyman, J., Pearson, E.: On the problem of the most efficient tests of statistical hypotheses. Phil. Trans. Roy. Soc. Ser. A **231**, 289–337 (1933)
7. Puri, M.L., Sen, P.K.: Nonparametric Methods in General Liner Models. John Wiley and Sons, New York (1985)
8. Seber, A.F., Lee, A.J.: Linear Regression Analysis. John Wiley and Sons (2003)
9. Shao, J.: Mathematical Statistics. Springer, New York (1999)

Chapter 9
Appendix D: Elements of Bayesian Statistics

The data from individual genetic experiments are rather noisy. Also, their large dimension requires the application of rather strict multiple testing corrections to reduce the number of false discoveries. This results in a relatively low power to detect important signals. On the other hand, there already exists a rich source of knowledge built upon years of research. Bayesian statistics naturally assists in the knowledge building process by integrating the information from current experiments with prior knowledge. In recent years, Bayesian statistics has gained popularity, due to the development of efficient algorithms based on Markov Chain Monte Carlo, which enable statistical inference for advanced hierarchical models and are very flexible with respect to modeling the uncertainty regarding prior information. In this section, we give a brief overview of the methods used in Bayesian statistics. More detailed information can be found in a multitude of books, e.g., [1–4] or [5].

9.1 Bayes Rule

Bayesian approaches enable us to integrate prior knowledge with the information given by new data. Let us denote the vector of observations by X. We mainly concentrate on a set of statistical problems where $X = (X_1, \ldots, X_n)$ is a random vector in $\mathbf{R}^n$, such that its distribution belongs to the class parametrized by the vector of parameters $\theta \in \Omega \subset \mathbf{R}^k$. However, the basic ideas of Bayesian statistics can be applied in much more complicated situations, e.g., the parameter θ belongs to an infinite dimensional class of density functions.

The prior information about the vector of parameters is coded in terms of a prior distribution with density/probability mass function $\pi(\theta)$. The information from the data is contained in the likelihood function $L(X|\theta)$, which is the conditional density/probability mass function of X, given the vector of parameters θ.

© Springer-Verlag London 2016
F. Frommlet et al., *Phenotypes and Genotypes*, Computational Biology 18,
DOI 10.1007/978-1-4471-5310-8_9

Now, using Bayes rule, we can calculate the posterior distribution of θ for given X as

$$\pi(\theta|X) = \frac{L(X|\theta)\pi(\theta)}{\int_\Omega L(X|\theta)\pi(\theta)d\theta}. \tag{9.1}$$

When applying a Bayesian approach, point estimators are usually obtained by minimizing some risk function, calculated with respect to the posterior distribution of θ. For example, minimizing the classical L_2 risk (corresponding to the mean square error), $L_2 = E[(\hat{\theta}-\theta)^2|X]$ leads to an estimate of θ equal to the mean of θ under the posterior distribution, i.e., $\hat{\theta} = E(\theta|X)$, while minimizing the L_1 risk (corresponding to the mean absolute error), $L_1 = E(|\hat{\theta} - \theta| \ |X)$ leads to an estimate of θ equal to the median of the posterior distribution, i.e., $\hat{\theta}$ satisfies $P(\theta \le \hat{\theta}|X) \ge 1/2$ and $P(\theta \ge \hat{\theta}|X) \ge 1/2$.

However, the posterior distribution also contains information about the uncertainty in the estimation of θ. In particular, based on the posterior distribution, one can calculate Bayesian credible intervals, which contain the real value of the unknown parameter with large probability. If $\theta \in R$, then the limits of the Bayesian credible interval for θ at the credibility level $1 - \alpha$ are usually determined by solving the following two equations:

$$\theta_{L,\alpha} = \max(\eta; P(\theta < \eta) \le \alpha/2) \quad \text{and} \quad \theta_{U,\alpha} = \min(\eta; P(\theta > \eta) \le \alpha/2). \tag{9.2}$$

9.2 Conjugate Priors

Often, it is not possible to derive the analytic form of the posterior distribution. Using classical Bayesian statistics, this problem is often solved by choosing the prior distribution in such a way that the posterior distribution belongs to the same class of distributions as the prior. This choice depends on the conditional distribution of the observations. Priors that satisfy this condition are called conjugate. Table 9.1 contains some popular examples of conjugate priors. Other examples of conjugate priors in classical inference problems can be found in [2] or at [6].

9.3 Markov Chain Monte Carlo

The recent increase in interest in Bayesian statistics has been triggered by the development of Markov Chain Monte Carlo methods. These methods allow us to estimate many Bayesian quantities by generating Markov Chains, whose limiting stationary distributions are equal to the required posterior distributions. This has led to a rapid expansion in the applications of Bayesian modeling by making it possible to use

Table 9.1 Examples of conjugate priors

Likelihood	Model parameters	Prior	Prior parameters	Posterior parameters
Binomial	p	Beta	α	$\alpha + \sum_{i=1}^{n} x_i$
			β	$\beta + \sum_{i=1}^{n} N_i - \sum_{i=1}^{n} x_i$
Negative binomial	p	Beta	α	$\alpha + \sum_{i=1}^{n} x_i$
			β	$\beta + rn$
Poisson	λ	Gamma	α	$\alpha + \sum_{i=1}^{n} x_i$
			β	$\beta + n$
Normal σ known	μ	Normal	μ_0	$\left(\frac{\mu_0}{\sigma_0^2} + \frac{\sum x_i}{\sigma^2} \right) / \left(\frac{1}{\sigma_0^2} + \frac{n}{\sigma^2} \right)$
			σ_0^2	$\left(\frac{1}{\sigma_0^2} + \frac{n}{\sigma^2} \right)$
Normal	μ	Normal	μ_0	$\frac{\nu\mu_0 + n\bar{x}}{\nu + n}$
			ν	$\nu + n$
	σ^2	Inverse	α	$\alpha + \frac{n}{2}$
		Gamma	β	$\beta + \frac{1}{2} \sum_{i=1}^{n} (x_i - \bar{x})^2 + \frac{n\nu}{n+\nu} \frac{(\bar{x}-\mu_0)^2}{2}$
Exponential	λ	Gamma	α	$\alpha + n$
			β	$\beta + \sum_{i=1}^{n} x_i$

very flexible hierarchical models, enabling us to model the uncertainty in the prior distribution. The most popular methods of generating Markov chains with a desired stationary distribution include the Gibbs sampler [7] and the Metropolis–Hastings algorithm [8, 9].

9.3.1 Gibbs Sampler

Consider the vector of parameters $\theta = (\theta_1, \ldots, \theta_k) \in R^k$. Let θ_{-i} denote the $k - 1$ dimensional vector $(\theta_1, \ldots, \theta_{i-1}, \theta_{i+1}, \ldots, \theta_k)$ and assume that for each $i \in \{1, \ldots, k\}$ it is known how to generate samples from the conditional distribution $\pi(\theta_i | X, \theta_{-i})$. Using the Gibbs sampler, new values of the vector of parameters are sampled by systematically drawing consecutive covariates from these conditional distributions. Let θ_{in} denote the value of θ_i in the nth step of the algorithm. The value of θ_i in the following, $n + 1$th, step is sampled from the conditional distribution $\pi(\theta_i | X, \theta_{1(n+1)}, \ldots, \theta_{(i-1)(n+1)}, \theta_{(i+1)n}, \ldots, \theta_{kn})$.

9.3.2 Metropolis–Hastings Algorithm

In many applications, the conditional distributions needed to apply the Gibbs sampler are difficult to calculate. In these situations, one can often use the Metropolis–Hastings algorithm, which only requires knowledge of the posterior distribution up to a multiplicative constant. Let θ_n denote the value of the vector of parameters in the nth step of the Metropolis-Hastings algorithm. The candidate for the new state $\theta_{n+1}^{\star}$ is sampled from some conditional distribution $q(\theta|\theta_n)$. Then $\theta_{n+1}^{\star}$ is accepted as the new state θ_{n+1} with probability

$$A = \min\left\{1, \frac{\pi(\theta_{n+1}^{\star}|X)q(\theta_n|\theta_{n+1}^{\star})}{\pi(\theta_n|X)q(\theta_{n+1}^{\star}|\theta_n)}\right\}.$$

In the case when the candidate state is not accepted, $\theta_{n+1} = \theta_n$.

When applying the Metropolis–Hastings algorithm, the proposal distribution $q(\cdot|\cdot)$ has to be chosen with care. When this distribution has a large variance, it allows us to jump to distant regions of the parameter space, which in principle should encourage mixing and increase the speed of convergence. However, this often leads to a low acceptance rate, which means that the Markov Chain often stays in the same state for extended periods of time. The optimal choice of the proposal distribution requires balancing these two counteracting tendencies and often requires some problem specific experience/knowledge.

For more detailed discussion on the rates of convergence of Markov Chain Monte Carlo methods to stationary distributions and issues regarding diagnostics, we refer the reader to [10, 11].

9.3.3 Hierarchical Models

Hierarchical Bayesian models enable us to model the uncertainty related to the choice of the prior distribution by using "second-order" priors on the prior parameters.

For illustration, we consider a classical model where $X_1, \ldots, X_n$ are independent random variables such that $X_i \sim N(\mu_i, 1)$. The aim is to estimate the vector of means $\mu = (\mu_1, \ldots, \mu_n)$. The classical maximum likelihood estimate of μ is simply

$$\hat{\mu}_{MLE} = (X_1, \ldots, X_n). \tag{9.3}$$

Using a Bayesian approach to estimate the μ_i, we assume that the prior distribution for each of the μ_i is $N(0, \sigma_0^2)$ (a conjugate distribution), which results in the following a posteriori distribution for the μ_i:

$$\mu_i|X_i \sim N(\tilde{\mu}_i, \tilde{\sigma}_i^2), \tag{9.4}$$

where $\tilde{\mu}_i = (1 - \frac{1}{\sigma_0^2+1})X_i$ and $\tilde{\sigma}_i^2 = \frac{\sigma_0^2+1}{\sigma_0^2}$ (see Table 9.1). Thus, minimizing the L_2 a posteriori risk leads to the following estimate of the vector of means

$$\hat{\mu}_{B1} = \left(1 - \frac{1}{\sigma_0^2+1}\right)\hat{\mu}_{MLE}, \tag{9.5}$$

which, compared to the classical maximum likelihood estimator, is shrunk towards zero. The amount of shrinkage depends on the variance σ_0^2 of the prior distribution.

The hyperparameter σ_0^2 of this prior distribution is responsible for the spread of the distribution of μ_i. When this parameter is unknown, information can be obtained from observed data. Using a fully Bayesian approach, one can use another layer of Bayesian modeling and impose a hyperprior distribution for this spread, $\pi(\sigma_0^2)$. In this case, one first calculates the joint posterior distribution of μ and σ_0^2

$$\pi(\mu, \sigma_0^2|X) \propto L(X|\mu)\pi(\mu|\sigma_0^2)\pi(\sigma_0^2),$$

where $X = (X_1, \ldots, X_n)$.

Then the posterior distribution of μ is calculated using

$$\pi(\mu|X) = \int \pi(\mu, \sigma_0^2|X)d\sigma_0^2. \tag{9.6}$$

Contrary to the posterior distribution of μ given by Eq. (9.4), the posterior distribution of μ_i resulting from (9.6) depends not only on X_i, but also on the whole vector X. This is due to the fact that the unknown hyperparameter σ_0^2 is estimated using the information on the spread of the elements of the X vector contained in the observations.

9.3.4 Parametric Empirical Bayes

Using an empirical Bayes approach, unknown hyperparameters are directly estimated based on the marginal distribution of the observed data.

For illustration, assume, as before, that $X_1, \ldots, X_n$ are independent variables and $X_i \sim N(\mu_i, 1)$. Moreover, assume that $\mu_1, \ldots, \mu_n$ are independent random variables from $N(0, \sigma_0^2)$. These assumptions imply that the marginal distribution of X_i is $N(0, \sigma_0^2 + 1)$.

Let us note that $||X||^2 = \sum_{i=1}^n X_i^2$. It is easy to check that under our assumptions $\frac{n-2}{||X||^2}$ is an unbiased estimator of $\frac{1}{\sigma_0^2+1}$. Thus, based on Eq. (9.5), the natural empirical Bayes estimator of the vector μ is given by

$$\hat{\mu}_{EB} = \left(1 - \frac{n-2}{||X||^2}\right)\hat{\mu}_{MLE}. \tag{9.7}$$

Interestingly, this empirical Bayes estimator coincides with the famous James–Stein estimator. For $n > 2$ and any given vector μ of true means, with respect to minimizing the classical mean square error, this estimator is better than the maximum likelihood estimator.

The hierarchical model discussed above can be naturally extended by including a location hyperparameter μ_0 (i.e., assuming that $\mu_i \sim N(\mu_0, \sigma_0^2)$ and that the variance of X_i is unknown, i.e., setting $X_i \sim N(\mu_i, \sigma^2)$).

More information on empirical Bayes methods can be found in [12].

9.4 Bayes Classifier

Consider the problem of classifying the sequence of i.i.d. observations $X_1, \ldots, X_n$ to one of k classes specified by the probability distributions $f_1(x), \ldots, f_k(x)$. For simplicity, we will assume that the loss for misclassifying an object from the ith class does not depend on the class to which it was wrongly classified. Let δ_i and π_i denote the loss for misclassifying an object from class i and the prior probability of being in class i, respectively, $i = 1, \ldots, k$. The procedure which minimizes the expected value of the loss function is called a naive Bayes classifier and selects the class for which

$$\pi_i \delta_i \prod_{j=1}^{n} f_i(X_j)$$

obtains its maximal value.

The Bayes classifier is often used in the context of pattern recognition (e.g., [13]).

References

1. Berger, J.O.: Statistical Decision Theory and Bayesian Analysis, 2nd edn. Springer, New York (1993)
2. Gelman, A., Carlin, J.B., Stern, H.S., Rubin, D.B.: Bayesian Data Analysis. Chapman and Hall, London (1995)
3. Ghosh, J.K., Delampady, M., Samanta, T.: An Introduction to Bayesian Analysis—Theory and Methods. Springer, New York (2006)
4. Gilks, W.R., Richardson, S., Spiegelhalter, D.J.: Markov Chain Monte Carlo in Practice. Chapman and Hall/CRC (1996)
5. Bolstad, W.M.: Understanding Computational Bayesian Statistics. John Wiley and Sons (2010)
6. http://en.wikipedia.org/wiki/Conjugate_prior
7. Geman, S., Geman, D.: Stochastic relaxation, gibbs distributions, and the bayesian restoration of images. IEEE Trans. Pattern Anal. Mach. Intell **6**, 721–741 (1984)
8. Metropolis, N., Rosenbluth, A.W., Rosenbluth, M.N.,Teller, A.H., Teller, E.: Equations of state calculations by fast computing machines. J. Chem. Phys. **21**(6): 1087–1092 (1953)

9. Hastings, W.K.: Monte Carlo Sampling methods using Markov chains and their applications. Biometrika **57**1): 97–109 (1970)
10. Mengersen, K.L., Tweedie, R.L.: Rates of convergence of the Hastings and Metropolis algorithms. Ann. Statist. **24**(1), 101–121 (1996)
11. Cowles, M.K., Carlin, B.P.: Markov chain Monte Carlo convergence diagnostics: a comparative review. J. Amer. Statist. Assoc. **91**(434), 883–904 (1996)
12. Casella, G.: An introduction to empirical Bayes data analysis. Am. Stat. **39**(2): 83–87 (1985)
13. Devroye, L., Gyorfi, L., Lugosi, G.: A Probabilistic Theory of Pattern Recognition. Springer, New York (1996)

Chapter 10
Appendix E: Other Statistical Methods

10.1 Principal Component Analysis

Principal component analysis (PCA) is a popular technique for reducing the dimensionality of data in multivariate analysis. Normally, the starting point is a set of m-dimensional data vectors $x_1, \ldots, x_n$. Define $X = (x_1, \ldots, x_n)$ to be the corresponding $n \times m$ data matrix. In this traditional setting, we have observations of m variables for n different individuals. In many cases, there will be an underlying correlation structure between the variables, which means that a smaller number of variables can be used to describe the underlying variation in the data. These variables are called latent variables and often have a natural interpretation.

One example of PCA in this book is given in Sect. 5.3.3, where m refers to the number of individuals and x_j contains all the SNP genotypes for individual j. Note that, in this example, the data matrix containing these genotypes is the transpose of the more usual form, where columns refer to SNPs and rows refer to individuals.

PCA can be performed either with centralized variables $x_j - \bar{x}_j$, or with standardized variables $\frac{x_j - \bar{x}_j}{s_j}$, where the definitions of the mean $\bar{x}_j$ and the empirical standard deviation s_j are given by Eq. (6.3). The more common approach is standardization, and for this presentation we will assume that the original data vectors have been standardized already.

Assume that $Z = (z_1, \ldots, z_m)$ is the random vector which generates the n data vectors, i.e., Z is described by the joint distribution of the m variables observed. Due to standardization, the (theoretical) correlation matrix, Σ can be computed as $\Sigma = E(Z'Z)$, all the diagonal elements of Σ are equal to one and σ_{ij}, the element in the ith row and jth column, is the coefficient of correlation between the ith and jth variables.

The aim of PCA is to find linear combinations of the data which exhibit the largest variation. Let the vector $t = (t_1, \ldots, t_m)$ characterize a general linear combination $Zt' = \sum_{i=1}^{m} z_i t_i$. It is assumed that the vector t is normalized, i.e., it is of length one. PCA is based on the eigenvalues and eigenvectors of the matrix $\hat{\Sigma}$, an estimator of

© Springer-Verlag London 2016

F. Frommlet et al., *Phenotypes and Genotypes*, Computational Biology 18,
DOI 10.1007/978-1-4471-5310-8_10

the correlation matrix. Let λ and t satisfy the equation $\hat{\Sigma}t = \lambda t$. From the specific properties of a correlation matrix, this equation has m solutions, where the ith solution corresponds to the (positive) eigenvalue λ_i and its corresponding eigenvector $t_{(i)}$.

We order the eigenvalues so that $\lambda_1 \geq \lambda_2 \cdots \geq \lambda_m > 0$. The proportion of the total variance of the data explained by the linear combination $Zt'_{(i)}$ is proportional to λ_i. Since the sum of the eigenvalues is equal to m, it follows that the linear combination $Zt'_{(i)}$ explains $\frac{100\lambda_i}{m}$ % of the total variation in the data. Based on these eigenvalues and eigenvectors, we can select k latent variables, where $k < m$, such that these k latent variables explain a very large proportion of the total variation in the data. One very simple rule of thumb for choosing the number of latent variables is to select those corresponding to eigenvalues greater than one, the mean value of the eigenvalues. For further details and methods of selecting the appropriate number of latent variables, see [2].

To see the intuition behind PCA, consider the following example: Suppose the variables we observe include height, weight, and waist size. These variables will be strongly positively correlated with each other. Hence, defining a linear combination of these three variables (which might correspond to an eigenvector of the correlation matrix where the coefficients associated with these three measurements are large), one obtains a single latent variable that can be interpreted as a general measure of size.

Now consider the case where the m variables and n cases correspond to m individuals and n genotypes, as described above. PCA would highlight not variables which are associated with each other, but individuals who are associated with each other. Hence, PCA can be used as a tool to investigate the structure of the population (see [3]).

10.2   The EM Algorithm

The expectation maximization (EM) algorithm is an iterative method for finding the maximum likelihood estimates of parameters describing a statistical model, when this model depends on unobserved latent variables. The EM algorithm alternates between performing an expectation (E) step, which calculates the expectation of the log-likelihood with respect to the conditional distribution of the latent variables, based on the current estimates of the parameters in the model, and a maximization (M) step, which re-estimates these parameters by maximizing the expected log-likelihood found in the E step.

Let us denote the observed data by $\mathbf{X}$ and the set of unobserved (latent) data by $\mathbf{Z}$. Our goal is to estimate the vector of parameters $\boldsymbol{\theta}$ by maximizing the marginal likelihood of the observed data, given by

$$L(\boldsymbol{\theta}; \mathbf{X}) = p(\mathbf{X}|\boldsymbol{\theta}) = \sum_{\mathbf{Z}} p(\mathbf{X}, \mathbf{Z}|\boldsymbol{\theta}). \qquad (10.1)$$

The sum on the right-hand side of Eq. (10.1) usually contains many components, which makes direct maximization of $L(\boldsymbol{\theta}; \mathbf{X})$ computationally intractable. Instead, the EM algorithm tries to find the maximum of the marginal likelihood by iteratively applying the following two steps:

(1) Expectation step (E step): Calculate the expected value of the log-likelihood function with respect to the conditional distribution of $\mathbf{Z}$ given $\mathbf{X}$ based on the current estimate of the parameters $\boldsymbol{\theta}^{(t)}$:

$$Q(\boldsymbol{\theta}|\boldsymbol{\theta}^{(t)}) = \mathrm{E}_{\mathbf{Z}|\mathbf{X},\boldsymbol{\theta}^{(t)}}\left[\log L(\boldsymbol{\theta}; \mathbf{X}, \mathbf{Z})\right]$$

(2) Maximization step (M step): Find the value of $\boldsymbol{\theta}$ that maximizes $Q(\boldsymbol{\theta}|\boldsymbol{\theta}^{(t)})$:

$$\boldsymbol{\theta}^{(t+1)} = \mathrm{argmax}_{\boldsymbol{\theta}}\, Q(\boldsymbol{\theta}|\boldsymbol{\theta}^{(t)}).$$

In [1, 4], it is shown that the sequence of values of the log-likelihood function obtained in this way monotonically approach a local maximum of the marginal likelihood.

References

1. Dempster, A.P., Laird, N.M., Rubin, D.B.: Maximum likelihood from incomplete data via EM algorithm. J. Roy. Stat. Soc. Ser. B **39**, 1–38 (1977)
2. Koch, I.: Analysis of Multivariate and High-dimensional Data. Cambridge University Press, Cambridge Series in Statistical and Probabilistic Mathematics (2013)
3. Price, A.L., et al.: Principal components analysis corrects for stratification in genome-wide association studies. Nat. Genet. **38**: 904–909 (2006)
4. Wu, C.F.J.: On the Convergence Properties of the EM Algorithm. Ann. Stat. **11**: 95–103 (1983)

Index

© Springer-Verlag London 2016
F. Frommlet et al., *Phenotypes and Genotypes*, Computational Biology 18,
DOI 10.1007/978-1-4471-5310-8

Printed in the United States
By Bookmasters